Innovation in Product Design

Monica Bordegoni · Caterina Rizzi
Editors

Innovation in Product Design

From CAD to Virtual Prototyping

 Springer

Monica Bordegoni
Dipartimento di Meccanica
Politecnico di Milano
Via G. La Masa, 1
20156 Milan
Italy
e-mail: monica.bordegoni@polimi.it

Caterina Rizzi
Dipartimento di Ingegneria Industriale
Università di Bergamo
Viale G. Marconi n.5
24044 Dalmine, BG
Italy
e-mail: caterina.rizzi@unibg.it

ISBN 978-0-85729-774-7   e-ISBN 978-0-85729-775-4
DOI 10.1007/978-0-85729-775-4
Springer London Dordrecht Heidelberg New York

British Library Cataloguing in Publication Data
A catalogue record for this book is available from the British Library

Cover design: eStudio Calamar S.L.

Printed on acid-free paper

Springer is part of Springer Science+Business Media (www.springer.com)

To Umberto

Preface

The advent of computers has changed the practice of engineering forever. No product is designed today without the use of computer-aided design (CAD) technology. This practice, introduced about 40 years ago, has resulted in more reliable products that are less expensive to produce and that are more attractive to potential customers. It has also changed the technical education and the practice of numerous professionals.

The idea for this book was born with the aim of giving an overview of the research fields and achievements on methods and tools for product design and innovation.

The introduction of computers in engineering is dated back to the 1960s, when computer-based graphic systems were developed with the aim of supporting engineering design, and improving the productivity of engineers. The importance and benefits were strongly recognized by large manufacturing companies, especially those in the automotive and aerospace industries.

In 1963 Ivan Sutherland developed his thesis entitled "Sketchpad: a Man-Machine Graphical Communication System", which can be considered to be the ancestor of modern CAD tools, as well as a major breakthrough in the development of computer graphics in general. In 1969 he also built the first Heads-up display at the University of Utah. He had the capability of envisaging the impact of this technology on design and engineering, and he can be considered the precursor of current Virtual Reality technologies. Since then, the technology has quickly evolved developing very sophisticated and performing visualization and rendering systems, as well as human computer interaction systems.

Since the early 1960s, computer graphics (CG) was attracting many people from the research field. At the beginning, CG and CAD were in some sense overlapping. In fact, CAD was the marketing application of CG technologies, and in many cases it became the synonym of 2-D or 3-D geometric modeling. Several theoretical studies were carried out not only by academic scientists, but also by industrialists. Just to mention few of them, S. Coons in 1967 introduced a method to describe free-form doubly curved surfaced of a very general kind. Automotive companies like Renault investigated on the mathematical definition of complex surfaces. In particular, in 1968 Bezier developed a specific kind of curves used for describing exterior car panels. In 1974 Requicha and Voelcher proposed a high-level description of products based on the constructive solid geometry—CSG approach. Since then, and for more than 30 years, the evolution of tools for product design has focused on the development of geometric modeling capabilities with the aim of reproducing complex shapes.

In the 1980s the research interest focused more and more on tools for supporting the product development activities, which were not solely addressing the product geometry. These tools aimed at not just documenting the result of the product design, but aimed at capturing how to reach the result. Their development was based on the consideration that the design process includes knowledge. In fact, knowledge is into the methodologies used, and into the decisions taken by the product engineers during the design. Therefore, following this new trend, the tools moved from being geometry-based to be parametric-based and feature-based. Features are intended as high-level entities that give a meaning to a set of geometric elements (e.g. faces, patches, ribs, pockets), which can be therefore manipulated directly without requiring a great skill and deep knowledge of the underlying mathematical models. The feature concept has been successfully adopted in mechanical engineering, as well-documented by Shah and Mäntylä in 1995. In the domain of aesthetic shapes, characterized by the great level of freedom that the designer has when defining product shapes, the so-called free-form features have been defined and widely studied. Free-form features are also intended to convey the emotions the users have when looking and touching shapes.

Geometric modeling started being handled by kernels who were separated from the CAD tools. Besides, the CAD vendors were not anymore the drivers of technologies development, but other sectors were taking their place, such as game and movie, and the military sectors.

Starting from the late 1980s, new technologies were introduced that were capable of elaborating complex problems and whose cost decreased more and more. Specifically, Personal Computers and workstations were introduced as a new and emerging market sector, and Microsoft and the NT operating system took hold.

In the context of product development, industry realized that managing the product geometry is only a part of the design problem. In fact, the management of the product information and of the product development process is crucial as well. Therefore, the attention was focused on engineering knowledge management, on modeling and simulation of processes, as well as on business process re-engineering (BPR).

Industry understood the importance of managing knowledge about the products. Engineering knowledge management has been introduced to bring knowledge about the product characteristics (materials, assembling, manufacturing, etc.) into the design process. Applications based on knowledge management allow us to design and configure a product, on the basis of a set of rules, which have been previously formalized and implemented into the application. Product configurators have been proposed with the aim to allow us to engineer a product by satisfying customer's requirements and standards of a specific domain, even if the product has never been developed in the past. The advent of knowledge-based engineering (KBE) applications profitably helped in simplifying and automating, even if not fully, the configuration process, both from a commercial and technical point of view. Product and process knowledge is structured, so as to bring an optimization of company's design processes, to maximize the reuse and sharing of company knowledge, and to integrate systems and documents within a unique application for products automatic configuration.

Thanks to the continuous increase of computational power and performances of computers, analysis and simulation practices started to be introduced into the product development process, as well as the use of the digital mock-up (DMU). The introduction of effective tools for computer-aided engineering (CAE) has affected the product design practice. In particular, the simulation is used in product design with a double function. Early in the initial design phases, it supports decision-making; whereas later it helps in validating the design with respect to specifications. Currently, the validation function is widely practiced in industry and this is where most investment is made.

The DMU was a full representation of a machine or of a system including all geometrical details. It was a sort of static representation used to check the collisions among the parts. Subsequently, the DMU has evolved into what we call today virtual prototyping, where the product geometry is enriched with product properties. Virtual prototypes are used to faithfully represent the "product-to-be", so as to be able to simulate its features, performances, functionality and usage before the real product is actually built. Virtual prototyping is more and more becoming a diffused practice in product development of various industrial sectors. Several research works have shown that virtual prototypes can be effectively used to validate the design solutions, already in the early phase of product design, when the engineering of the product is also in the early phase or even not started. This practice can also be used for checking the correspondence of the concept design with the user's needs, and also for checking the users' acceptance of new products through tests performed directly with end users.

For many years, researchers have worked on visualization issues, and the resulting tools are very good in terms of rendering the visual effects of the products they can provide. Shades, reflections and textures contribute in enriching the product appearance. Physical prototypes have traditionally been a common way of representing a product design: physical models are produced, often in a real scale, to evaluate a product by exploring it visually and haptically. In the last decade, the interaction with virtual objects has been extended with the possibility of touching the object surface and of physically handling its parts, which is practically implemented through haptic devices. Today, thanks to the current advances of the technology, the implementation of virtual prototyping applications can be based on virtual reality technologies, where products and environments are fully virtual, or on augmented or mixed reality technologies, where the product and the environment consist of a mix of virtual and real components.

Starting from the 1990s the attention of industry has moved to a more strategic vision and to the idea that it is not important specifying what has to be manufactured and produced, but the product life cycle is also important from the marketing point of view. The new technologies are applied to increasingly complex projects. This means managing massive amounts of design data. PLM—product lifecycle management tools have been introduced with the aim of increasing the industrial productivity. PLM manages all the phases of product design, from conception, to manufacturing, service and disposal, and integrates people, data, processes and business systems.

PLM systems are extending their domain of application upward the preliminary phases of design and by embedding more abstract representations of the product, but still they are far from systematizing inventive design phases and the link between the development of a conceptual solution and the definition of the product geometry. Recently, computer-aided innovation (CAI) systems have started addressing these lacks, but the domain borders of this emerging technology are still fuzzy and in any case CAI systems suffer of limited interoperability with downstream computer-aided tools.

Virtualization and knowledge management have been recognized as two mega-trends for the period to come. This book presents a variety of research topics related to product development and innovation from experts in many different fields. Chapter 1 includes an overview of the history of tools for product design that evolved in the current tools for the whole product life cycle management. Chapter 2 deals with new and more effective methods for conceptual and embodiment design of products, which are based on the automation of embodiment design tasks. Chapter 3 aims at highlighting the role of knowledge as a key enabler for effective engineering activities in the light of emerging enterprise collaboration models. An overview and research issues on methodologies and tools for knowledge management of products are presented and discussed in Chaps. 4 and 5. In particular, Chap. 4 introduces principles and tools for design automation in mechanical engineering and methodologies for real industrial applications, while Chap. 5 discusses about product knowledge of the industrial design domain that is formalized into form-features. Chapter 6 describes some aspects of how the evolution of computer-aided engineering has affected the product design practice. The subject is treated from an industrial point of view, where some examples typical of the aeronautical field are given. Chapter 7 deals with methods, tools and issues related to the practice of Virtual Prototyping used in the product development process. Chapter 8 presents the state-of-the-art in the domain of digital human models used for product validation and assessments. Finally, Chap. 9 deals with the generation of the physical models of products, and discusses about how the physical prototyping process influences the product design.

We hope the reader will enjoy and learn about the history, the evolution and the future perspective of research in the domain of methods and tools for product design and innovation.

Milan, February 2011   Monica Bordegoni
Bergamo, February 2011   Caterina Rizzi

Acknowledgment

All the authors of this book are grateful to Umberto Cugini, for his guidance teachings, for his continuous inspiring ideas and vision about the future of methods and tools for product development, with which he has directed our research and careers.

Contents

Chapter 1
The Evolution of Digital Tools
for Product Design

Massimo Fucci

Abstract The development of digital tools for product design started with the research works on computer graphics in the 1960s. By the end of the decade a number of companies were founded to commercialize the first CAD programs. Since then, the tools and the technology used in product development have advanced rapidly, were based on different modeling methods, and used in several industrial sectors. During the years, the developers have concentrated less on the functionality of the design tools, and more on issues related to the management of the product life cycle. The chapter tells the story of the digital tools for product design, by addressing their evolution till the current product life cycle management solutions that integrate various tools for better integrating the product development phases.

1.1 Introduction

Modern engineering design and drafting started with the development of descriptive geometry. Drafting methods improved with the introduction of drafting machines, but the creation of engineering drawings changed very little until the beginning of the 1960s, when there were a number of researchers working on the idea of computer graphics for design.

Considerable studies were carried out on the development of real-time computing. In particular, Ivan Sutherland worked on his Ph.D. thesis at the Massachusetts Institute of Technology in 1963, which was titled "Sketchpad: A Man–Machine Graphical Communications System" [1]. The idea was about a system letting

M. Fucci (✉)
Pentaconsulting S.r.L., Piazza Caiazzo 2, Milan, Italy
e-mail: massimo.fucci@pentaconsulting.it

M. Bordegoni and C. Rizzi (eds.), *Innovation in Product Design*,
DOI: 10.1007/978-0-85729-775-4_1, © Springer-Verlag London Limited 2011

designers use a light pen to create engineering drawings directly on a CRT monitor. The drawings could also be manipulated, duplicated and stored. Sketchpad contributed to the advent of graphic computing and included such features as computer memory to store drawn objects, rubber-banding for simpler line construction, the ability to zoom in or out on a display, and techniques for making perfect lines, corners and joints. It was the first example of graphic user interface (GUI). Besides, Sutherland also started developing the first algorithms for hidden lines removal in 3D drawings, essential to generate realistic renderings for CAD models.

In the same period, parallel work was being done at the General Motors Research Laboratories. Here, Patrick Hanratty developed the first commercially available software for mechanical drafting.

In the 1960s we assisted also to an advance of hardware technologies. In particular, it was also developed the first digitizer (from Auto-trol) and DAC-1, the first production interactive graphics manufacturing system. By the end of the decade, a number of companies were founded to commercialize the first CAD programs, including SDRC, Evans & Sutherland, Applicon, Computervision and M&S Computing.

Since then, the tools and the technology used in product development have advanced rapidly [2]. The following sections deal with the evolution of product modeling and the evolution of tools for product design, with the issues related to the integration of the product development phases, and the solutions proposed by modern systems for the management of the product life cycle. Then, we report some case studies that demonstrate the potentiality and benefits of the use of these new tools.

1.2 The Evolution of Modeling

The way in which the product representation in its digital form has evolved can be described through a sequence of levels (Fig. 1.1). At each level the product is represented through a model, starting from a very simple one, and growing up in complexity. The quantity of formalized information embedded into the model grows with the progressive evolution of the model.

The levels represent the different families of product modeling systems that were initially based on 2D geometry and then evolved into wireframe model, boundary model, solid model, parametric and functional model and finally to DMU (Digital Mock-Up and) and to VP (Virtual Prototyping).

In the course of the evolution, at each level, explicit items of information have been added into the system (right-hand side of the diagram shown in Fig. 1.1). At each level, the system needed to be complemented by a certain amount of knowledge that the product designers are required to have in order to reach the design goal.

In the seventies, CAD systems were essentially modelers and editors of documents. These documents, i.e. engineering drawings, were the common way to

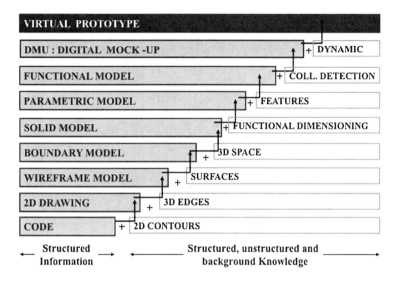

Fig. 1.1 Modeling levels (courtesy of Prof. Umberto Cugini, Politecnico di Milano)

specify the results of the design process and to communicate information in a complete, non-ambiguous and standardized way.

In the eighties, CAD systems were more oriented to the generation and use of three-dimensional digital models based on solids and surfaces. This approach has brought a fundamental change in the way of managing product designs: the difference between models and their representations. Models are simplified geometrical description of physical products, and engineering drawings are representations derived from digital models.

In the nineties, feature modeling was introduced in CAD systems with the aim of overcoming the strong limitations related to simple and pure geometric models used for representing complex products. Features were used to add knowledge about the various contexts and design phases. This knowledge concerned functional, manufacturing, assembling and measuring aspects. The revolutionary capability of feature-based CAD systems was capturing and modeling, with the aim of reusing, the product definition process as a way to represent "how" the design was carried out.

At the top of the diagram there are the DMU, which is basically a static representation of the product, mostly limited in the interaction with users, and the VP, which ideally should be used in the same way as the physical prototype. Virtual Prototyping is growing as a technology separated from the traditional CAD tools, since it is a different way of representing a product and cannot be an incremental development built on CAD tools.

Recently, we assist to an evolution from simple and static simulations to a new approach that is not based on the modeling of objects, but on the modeling of the physical phenomena. The shape is important, but it is just one of the attributes of the product and it is not static. In fact, the shape is also a function of the environment, of time, of the history of the phenomena affecting the object. Actually, this is a big shift in the mentality of designers, and also in the approach that is used to develop the products. In this new perspective, products and systems are modeled by means of physics-based tools, which are based on simulation of the products behavior, and on simulation of their interaction with the environment. The product being designed has to be checked during the interaction with other machines, with the environment, and in a context where various people are involved in the interaction with it—i.e. users, but also people involved in the manufacturing and in the assembly process.

This paradigm shift is from a geometrical view to a functional view. The product description is provided in terms of a very detailed shape, and also of many other aspects that include aesthetics, functional, ergonomics and that also deal with manufacturing and assembling.

1.3 The Evolution of Tools for Design

By the 1970s, research had moved from 2D–3D. Major milestones included the development of Bezier, B-splines and NURBS representations [3–6], which formed the basis of modern 3D curve and surface modeling, and the development of the PADL (Part and Assembly Description Language) solid modeler [7].

With the emergence of UNIX workstations in the early 1980s, commercial CAD systems like CATIA and others began to be used, especially in aerospace and automotive industries. But it was the introduction of the first IBM PC in 1981 that determined a large-scale adoption of CAD in industry. In 1983 the new company Autodesk, founded by a group of programmers, released AutoCAD, the first significant CAD program for the IBM PC. AutoCAD is a significant milestone in the evolution of CAD. The tool offered as much as the same functionality of the other CAD tools, at a significant lower cost. From then, increasingly advanced drafting and engineering functionality of design tools became more affordable. But it was still largely 2D.

The situation changed in 1987 with the release of Pro/ENGINEER, a CAD program based on solid geometry and feature-based parametric techniques for defining parts and assemblies. Pro/ENGINEER ran on UNIX workstations, since the PCs of that time were not powerful enough. The introduction of this technology, including the parametric modeling approach, sets a change in the way of designing products. At the end of the 1980s, several 3D modeling kernels were released, most notably ACIS and Parasolids, which formed the basis for other history-based parametric CAD programs.

Since then, the power of PCs grew up significantly, so that by the 1990s they were capable of performing the computations required by 3D CAD. In 1995 SolidWorks was released, which was the first significant solid modeler for Windows. This was followed by Solid Edge, Inventor and many others.

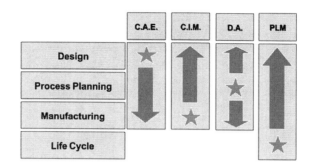

Fig. 1.2 Integration of the product development phases (courtesy of Prof. Umberto Cugini, Politecnico di Milano)

In the meantime, KBE (Knowledge Based Engineering) tools appeared on the market to support Design Automation; they evolved from initial elementary implementation to the actual one characterized by simple programming language, powerful tools to define customized GUI and to integrate external programs (CAD systems, FE solvers, spreadsheet, data base, etc.) [8]. In the last decade, many of the original CAD developers from the 1960s have been acquired by newer companies. Currently, we assist to a consolidation of the CAD tool providers into four main players, namely Autodesk, Dassault Systèmes, PTC and UGS (now Siemens PLM software), along with a group of smaller developers.

The transition from 2D to 3D modeling took longer than the previous transition from the use of drawing board to 2D digital modeling. This was mainly due to the fact that the conception of a product and its representation in 3D was not a technological problem, but a new way of thinking that engineers had to assume. And this was a sort of barrier that started pointing out that human resources are critical factors of success in companies.

The modern CAD era has been subjected to improvements in modeling, incorporation of analysis and management of the products we create, from conception and engineering to manufacturing, sales, maintenance and dismantling: this is today known as PLM (Product Lifecycle Management).

1.4 Integration of the Product Development Phases

Traditionally, there has been an attempt to integrate the three main phases typical of product development: design, process planning and manufacturing (Fig. 1.2).

CAE (Computer Aided Engineering) was a term used to indicate this integration, where the focus was on design. It referred to the engineering of the design, where there was an automatic detailing of what was designed by the designers.

CIM (Computer Integrated Manufacturing) referred to the mass production environments, where typically there is not a large innovation on design, but the focus is on the reduction of the costs of the manufacturing process. This typically applies to the automotive field, to household appliances, etc., where the difference in the success of the products is given by the innovation of the manufacturing process aiming at reducing the costs and time. And the design is constrained by the new manufacturing process.

In the electronics field, typically a complementary approach is adopted, which is named DA (Design Automation). Depending on the amount of device that we are supposed to produce, we choose the technology, and depending on that there are well-defined rules for the design and also for the manufacturing. There are some automatic compilers where one specifies the requirements, chooses the technology, and the design and manufacturing are automatically defined.

More recently, there has been an attention to a more inclusive integration of tools for product development. This integration is named PLM.

Various definitions have been proposed to explain and synthesize the PLM concept [9–11]. PLM system can be seen as a set of tools and methodologies to manage the evolution of a product during its life cycle from its conception to the disposal. It is not only a mere introduction of new technologies within a company, but it is more a new organizational paradigm and a business strategy to enhance integration and collaboration during definition, share and use of engineering data, i.e. all the information needed along the product life cycle. It implies the coordination and integration of processes and applications used to define and manage the so-called virtual product with those to manufacture and maintain the physical product. Therefore, it can be considered the glue to connect such environments [12–14].

1.5 The Management of the Product Life Cycle

At the beginning of the new century, the engineers and designers were being asked to create more, faster and with higher quality. At that time, CAD tools had not really changed much beyond adding more features and updating the user interface. There was the necessity to handle product data in a more systematic and coherent way, among the various phases of product development process. Besides, companies started having design offices geographically distributed, with the consequent necessity to manage resources over the network. Therefore, the integration of design activities, interoperability of data and processes and web tools were some of the major key issues.

Alibre by Alibre Design was the first 3D CAD software able to perform client–server 3D modeling over the Internet and, taking it as an example, other Web-enabled CAD tools were developed. Since middle of 1990s, Toyota Caelum's TeamCAD permitted 3D modeling across LANs. In mid 2000, AutoCAD 2000i, a Web-enabled CAD system, appeared on the market. Engineering drawings could be viewed with a Web browser and some online simple collaboration was possible using Microsoft Net Meeting.

In late 2000, there were successful stories demonstrating that the integration of CAD software, PDM software and the Internet gave engineers and designers the ability to view and collaborate on a single digital "master". This allowed companies to save in time and travel expense, but also eliminated the traditional mismatch problems inherent in the design and production of a complex product, due to a globally dispersed team. This permitted to visualize complex 3D models and to manage the complexity of large products assemblies as those of automotive and aerospace industry, and significantly reduce product development times.

In such a context, technology vendors had to demonstrate the effective capability of their solutions to manage the process flow and the engineering data, of which 3D CAD models represent more and more a small percentage.

The origin of PLM can be found at the beginning of 1990s as university research mainly concerning manufacturing databases. Only at the end of 1990, PLM became popular in industry. In parallel 3D CAD technology vendors redefined themselves as "PLM solution provider", and the four major vendors– Dassault Systèmes, Parametric Technology, UGS and SDRC-changed, accordingly, their corporate images, marketing and sales processes. Today, three PLM solution vendors (IBM-Dassault Systèmes, UGS-Siemens PLM Solutions, and PTC) and Autodesk are dominating this market area.

1.6 Case History

The adoption of 3D modeling as well as of PLM solutions have found initial obstacles in companies. In fact, technology providers have pushed the concepts mainly from a marketing point of view. A survey on a sample of 1,000 companies has shown that CAx (Computer Aided x) tools have been adopted by most of the companies, especially CAD ones, while PDM/PLM solutions are less diffused in SMEs (Small Medium Enterprise). Main reason relies on management and organizational aspects and the capability to understand and manage continuous technology and market changes.

Companies needed convincing examples of successful applications of PLM solutions. Hereafter, it is reported a successful example of an association of industrial companies, which has promoted the use of these technologies, also through the support in the development of best practices, and their dissemination through national media.

In 1999, the Italian association SoluzionImpresa® (http://www.soluzionimpresa.it) launched an initiative involving hardware and software technology providers, some Italian universities, and industrial associations and categories, with the aim of proposing solutions to problems stated by the companies, independently from specific tool brands.

The initiative, conducted over the years, demonstrated that companies can remain competitive only if a continuous process of systematic innovation is adopted. The most critical factor of success is not related to technological solutions but rather to the capability of developing a collaborative environment that supports an effective exchange of product information, also through cross modality, through the all company processes and departments, and more and more with the involvement of a network of external suppliers.

The following sections present some selected industrial cases where PLM solutions have been successfully implemented, with clear benefits for the companies.

1.6.1 Case I: Integration of Design, Simulation and Data Management

The market of electrical equipment for residential use has been greatly modified: home is a place now characterized by functionality, liveability, security and, not at least, harmoniousness and pleasantness. It is the ability to satisfy constantly new customers' needs over time that has made Vimar (http://www.vimar.eu) a leading provider of electrical equipment for consumer, healthcare and tertiary markets. The company was founded in 1945 and demonstrated immediately to be a production leader of "pear switches". The business has continued successfully through the introduction of some modular series, but it was with the advent of the concept of domotic products, that began the era of dedicated solutions, as for example monitoring of intruders or accidental events, house wellness and comfort through temperature control.

Nowadays, Vimar company develops its production in nine civil modular families, five of which have been recently designed and introduced, each consisting of about 1,000 components. The technical department, in order to cope with the market requests, has implemented a product development process able to maintain high reliability for the entire products range.

The technical department has faced a number of important changes in the design process, moving from wood-made prototyping to bi-dimensional design and, finally, to tri-dimensional design. This evolution has allowed the company to introduce, since 1999, the use of 3D CAD by means of which all the products are designed as modular and result from a top down design approach, which meets all the technical specifications defined by the reference standards and by the corporate know-how. The parametric and associative functionality of the CAD tool fully satisfies the design needs, by facilitating the tasks of the design team, which has to design a rich family of products and maintaining all the compatibility features required by the entire products line. In fact, Vimar company catalogue includes a total of about 7,000 product codes, where the products need to interact with each other, by respecting the compatibility and functionality rules.

This is a non-trivial problem for the technical department of the company, which has to assure compatibility and quality of the 3D CAD components, while testing the reliability criteria of the single parts through the use of CAE simulations (finite elements analysis, rheological analysis, dynamic analysis, etc.).

Fig. 1.3 Example of a
modular product
(courtesy of Vimar)

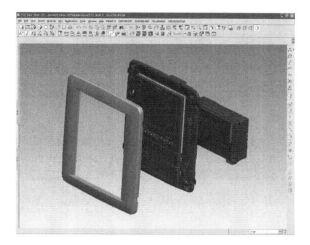

In the market context of these kinds of products, technicians, who were in charge of installing the products, mainly made the purchasing choices. Each choice was made on the basis of product modularity and the time reliability. Today, customers intervene in the choice, and they mainly decide on the basis of product aesthetics and style. Consequently, the design should take care of the product functionality and performance, as well as aesthetic features.

Vimar has demonstrated to be able to cope with these requests. The results can be seen in the EIKON series (Fig. 1.3), where the selection of the switch colors has been made so as to better highlight the precious materials the plates are made of (natural stones, crystals, woods, stainless steel and metals painted with ecological materials).

To manage the design process and data of the products, equipments and machinery, the company has adopted a PDM (Product Data Management) solution which provides an environment to share the data among the all stakeholders and company departments involved in the product development.

Thanks to the effective use of the adopted solutions, Vimar company has achieved several objectives, including costs reduction and a significant decrease of the time spent to design and develop the different products families. In particular, there was a reduction of about 30% of the time spent for the development of a new product family (consisting of about 1,400 products) and a significant reduction of the raw materials, without affecting structural performances of many back boxes listed in the catalogue.

1.6.2 Case II: A PLM Solution for Garden Machine

In order to efficiently compete and excel in a rapidly evolving market, a key strategy is the adoption and the maintenance over time of high quality standards.

Fig. 1.4 Example of a
modular product (courtesy
of Gianni Ferrari BFB)

Fig. 1.4 Example of a modular product (courtesy of Gianni Ferrari BFB)

The Gianni Ferrari BFB group (www.gianniferrari.com) is an example. The company is located in Italy in one of the most advanced technological districts for mechanical and hydraulic sectors. It is well known for its highly professional garden machines and characterized by a strong attention to innovation and production of high-quality products.

The company has been the first one to put on market innovative products, such as the "zero turning radius" (Fig. 1.4) mower machines. It is a successful product thanks to a better performance during operation: the container facilitates the material expulsion and the mower is equipped by floating bladed discs with inverse rotation.

The Gianni Ferrari BFB group develops a dozen families of garden machines for professional or multi-purpose use, thanks to a set of accessories that can be customized for specific uses, such as the cleaner, the earth moving equipment, the fork lifts and the snowplows. All the machines can be personalized and the product configuration can be easily described through a BOM (Bill Of Materials), up to ten levels.

The development and management of a so articulated production system requires the use of a series of applicative software able to support the product development cycle and to ensure an effective and efficient management of product data through the design phases, both inside and outside the technical office.

In order to support adequately the product development cycle, the company has adopted a 3D CAD solution for product design, a CAE solution for the product analysis and simulation, and a PDM solution for the management of engineering data.

The 3D CAD tool is mainly used to develop and manage complex 3D models, constituted by sub-assemblies that are able to represent digitally the real products. This has allowed the company to avoid the presence of inconsistencies since the design phase and to reduce-for some product models-by 3–6 months the time of the development cycle, compared to the canonical 18–24 months that were necessary before the introduction of a complete 3D CAD-based approach.

The structural analysis and simulation performed on the digital model, carried out by a CAE module integrated with the CAD tool, allow the designers to modify in real-time the characteristics and features of the final product. The simulation and analysis features have allowed the company to develop a new mower model that is more performing and about 40% lighter than the old one, with a consequent significant reduction of the production costs in terms of materials and times.

In order to have a real competitive advantage, the product data created during the design process have to be easily accessible, reliable and consistent, as well as they have to be used (and re-used) primarily by the technical office but also by all the other departments involved in the product development.

The PDM solution adopted was installed both in the technical office and in the sale office, with the aim to facilitate and speed up the prototyping phase of the new products, and to guarantee the consistency among the technical requirements. The use of the PDM has also allowed the company to generate a standard workflow that greatly facilitates the synchronization of the several process phases, by giving the possibility to maintain the control over the production, and especially by reducing hidden costs.

1.7 The Benefits of Integration

The analysis of the successful case histories permits to understand the key factors and the best practices associated with the adoption of a PLM solution and, at the same time, comprehend the simultaneous presence of two factors: one of technical nature and another one related to management, especially to human resources. The experience has demonstrated that the second aspect overwrites the first one.

First, the real benefits derived from the implementation of a PLM solution, and their increase over time, are possible only with the adoption of a 3D-based design approach to define geometric, topological and functional features of new products in a virtual environment. This is the only possible strategy to support properly technical and management decisions with regard to times, costs and product functions.

A second consideration, emphasized in the reported cases, is about the benefits of the software applications. Independently from the specific adopted CAD/CAE systems (i.e. from the commercial brand of the software tool), the benefits for the technical departments and for the overall company can be substantial only if the company installs/implements—maybe in different periods—the SW tools, which support the various phases of the product development process and if an effective integration within the whole organization is realized. In other words, we cannot expect an increase of the benefits only on the basis of the improvement of the 3D CAD features, but it is necessary to introduce other modules, i.e. the simulation modules (CAE), the Product Data Management (PDM) modules and, not least, the modules for data integration.

The third consideration is related to the fact that once the various modules (CAD, CAE, PDM) are implemented, the impact is not only technical but also

Fig. 1.5 PLM benefits

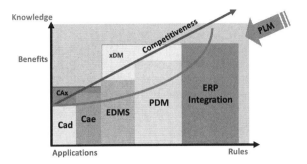

organizational. Therefore, good results are achieved only if the company management has a clear vision about the objectives and the steps necessary for making operative and effective a collaborative environment, whose activities are traceable to grant an adequate response time to the market requests.

The fourth consideration concerns the importance of the product style. In fact, currently the aesthetic impact has a major importance in the choice of the buyer also in the industrial field, and not only in the consumers' domain. The machine tools of the last generation are an example. Actually, they have not a square shape anymore, and their grey appearance has given the way to colors. As a matter of fact, when one visits a workshop, the impact is fully different from the one of several years ago. The style becomes important also for the details that are often hidden. For example, cabinets mounted in barely accessible positions and especially not visible, are often chosen by the customers or by the production process engineers on the basis of their appearance, besides the necessary technical data.

The diagram reporting the benefits vs. applications (Fig. 1.5) shows both the performance of benefits and competitiveness. This is a strategy, which represents the foundations for capturing, formalizing and spreading know-how in terms of product and processes within the company. It has been increasingly proved that this approach is an element of success, as it guarantees traceability, re-usability and adequate time-to-market in order to face rapidly changes of the market needs.

Finally, analysing the market, it is not sufficient to be excellent at the technical level—it is obvious that ideas and intuition will pay over time—but it is also necessary to excel at human and management level. Only in this case it is possible to develop the right product, "useful and with a nice appearance" within the correct time-to-market, with the right cost for the potential buyers, and, at the same time, the right margins for the manufacturer.

1.8 Conclusions

This chapter has presented a short overview of the evolution of the digital tools for product design. Recently Computer Aided Design tools have been integrated into more comprehensive environments, named PLM, which manage the whole product life cycle. "PLM solutions", as they are called, are environments that unify the technical information system with the management information system, where the information is treated in a coherent and unambiguous way, in all phases of the product development. Several PLM implementations have demonstrated the capability of reducing the time-to-market and the costs, and of increasing the competitive capability of the companies.

In a future perspective, we can identify some technologies that will potentially have greatest impact on the future of PLM/CAD tools.

First there is the necessity of developing new, very friendly, very interactive interfaces, which will attract new users and support faster design iterations. Today, we can build digital models of all shapes but it is still too hard way. Actually, it is not a question of whether we are able or not to model shape but it is more a question of whether all users are able or not to model any shape. Therefore, we expect a significant improvement of human computer interfaces, where novel technology will allow us to see 3D models through stereoscopic Head Mounted Display or through holographic representations, and it will also allow us to interact with digital models through new modalities, like Brain Computer Interfaces and haptic interfaces.

Another important issue concerns the possibility of embedding the simulation tools into the design tools, so as to enable designers to analyze design data as they model, and immediately know the result of any change before they commit to it. Interoperability of data and processes is equally important. The possibility of working with a unique and consistent digital model all over the product development process, independently from the software applications used, would be certainly beneficial in terms of productivity and quality of the results.

The computation power offered by new hardware technology, as GPU and multi-core systems, will allow us more and more to manage large digital models, to perform complex analysis and rendering, in real-time.

References

1. Sutherland I (1963) Sketchpad: a man-machine graphical communication system. PhD thesis, Lincoln laboratory, Massachusetts institute of technology, Cambridge
2. Lee K (1999) Principles of CAD/CAM/CAE systems. Addison-Wesley Publishing, Reading, MA, USA
3. Bezier PE (1968) Procédé de définition numérique des courbes et surfaces non mathématiques. Automatisme XIII 5:189–196
4. Farin GE (1997) Curves and surfaces for computer-aided geometric design: a practical guide, 4th edn. Academic Press, San Diego, CA, USA
5. Coons SA (1997) Surfaces for computer-aided design of space forms. technical report. Massachusetts institute of technology, Cambridge, MA
6. Mortenson ME (1997) Geometric modeling, 2nd edn. Wiley, New York
7. Requicha A, Voelcher H (1974) PADL part and assembly description language. Production automation product, University of Rochester, Nov 1974
8. Colombo G, Pugliese D, Rizzi C (2008) Developing DA applications in SMEs industrial context. In: Cascini G (ed) Computer aided innovation (CAI), WCC 2008, Milan, Sept 7–10, 2008. Springer, London

9. CIMdata (2002) Product life-cycle management. Empowering the future of business. CIMdata Report. http://www.cimdata.com/publications/pdf/PLM_Definition_0210.pdf. Accessed Feb 2011
10. Bacheldor B, Kontzer T (2003) What PLM is and is not. http://www.informationweek.com/news/software/productivity_apps/showArticle.jhtml?articleID=12800330. Accessed Feb 2011
11. Datamation limited (2002) Understanding product lifecycle management datamation limited report n. PLM-11, rev. 1.0, Sept 2002. http://www.purdue.edu/discoverypark/PLM/SME/Understanding_PLM.pdf. Accessed Feb 2011
12. Ramelli A, Rizzi C, Ugolotti M (2004) PLM paradigm in SMEs. In: Proceeding of the 14th international CIRP Design seminar 2004, Design in the global village, Cairo, 16–18 May 2004
13. Bertoni M, Bordegoni M, Cugini U, Regazzoni D, Rizzi C (2009) PLM paradigm: How to lead BPR within the product development field. comput ind. doi: 10.1016/j.compind.2009.02.004
14. Cugini U, Bordegoni M (2005) From systematic innovation to PLM. In: Dankwort W (ed) Holistic product development—challenges in interoperable processes methods and tools. Shaker-Verlag, Aachen

Chapter 2
From Computer-Aided (Detailed) Design to Automatic Topology and Shape Generation

Gaetano Cascini and Federico Rotini

Abstract This chapter surveys the evolution of Computer-Aided systems in terms of support to the earliest stages of design and more specifically to the embodiment design phase, when functional requirements and related structural and manufacturing constraints must be translated into a working solution, i.e., the generation of topology and shape of a mechanical part. After an introductory discussion about the context and the limitations of current systems, the chapter summarizes the research outcomes of two projects: the first, namely PROSIT (From Systematic Innovation to Integrated Product Development), aimed at bridging systematic innovation practices and Computer-Aided Innovation (CAI) tools with Product Lifecycle Management (PLM) systems, by means of Design Optimization tools. The second, coordinated by the authors, is a prosecution of PROSIT and proposes the hybriDizAtion of Mono-Objective optimizations (DAeMON) as a strategy for automatic topology and shape generation. The latter is clarified by means of two exemplary applications, one related to a literature example about Genetic Algorithms applied to multi-objective optimization, the second to an industrial case study from the motor-scooter sector.

G. Cascini (✉)
Dipartimento di Meccanica, Politecnico di Milano, Via La Masa 1, 20156 Milano, Italy
e-mail: gaetano.cascini@polimi.it

F. Rotini
Dipartimento di Meccanica e Tecnologie Ind.li, Università degli Studi di Firenze,
Via S.Marta 3, 50139 Firenze, Italy
e-mail: federico.rotini@unifi.it

M. Bordegoni and C. Rizzi (eds.), *Innovation in Product Design*,
DOI: 10.1007/978-0-85729-775-4_2, © Springer-Verlag London Limited 2011

2.1 Introduction

The last century has seen the development of more and more structured methods and procedures to support the Product Development cycle, both in terms of techniques to guide designers' decisions and of technologies to aid analysis and synthesis tasks. Three main ages are recognized in the literature [1]: the era of productivity characterized by an increase of demand by society for the acquisition of technical objects and consequently the focus on productivity improvement and costs reduction; the era of quality characterized by the necessity for rigorous steps of measurement and monitoring the production in order to increase the profitability, towards a total strategy of optimization of its efficiency; the era of innovation characterized by the need for structuring not only productivity and total quality, but also building a strategy of systematic innovation to bring in the market products addressing new users' needs or new ways to satisfy already identified needs and requirements.

The first two ages have firstly involved the optimization of the production departments (both methods and technologies) in order to reduce the unitary cost of a product, i.e., adopting lean production approaches, and to guarantee its quality, i.e., ensuring the robustness of the related manufacturing processes.

More recently, the focus has been switched to the engineering design tasks not only because they dramatically impact costs and quality, but also due to the emergence of innovation as the key for being competitive in the global market.

Despite methods and tools for engineering design have radically evolved in the last decades thanks also to the availability of computational resources not comparable with any human effort, the engineering design process can still be considered as a series of three major stages: conceptual design, embodiment design and detailed design [2].

Pahl and Beitz consider conceptual design as "a search across an ill-defined space of possible solutions, using fuzzy objective functions and vague concepts of the structure of the final solution''. According to this classification, embodiment design operates with a selected (during the conceptual design stage) initial design configuration and aims to further specify the subsets form in the whole system.

Nevertheless, in order to be competitive in current markets, companies must combine the capability to propose innovative products and services with efficient development processes. In this perspective, the authors think that the vision of Pahl and Beitz on conceptual design needs to be updated, since a proper identification of the design goal, as well as a formalization of the project constraints are necessary to reduce, from the very beginning of the innovation process, waste of time and resources through useless trials and errors.

Besides, the efficacy and the efficiency of the innovation process are highly impacted also by the adoption of suitable methods and tools in the embodiment design stage, i.e., that part of the design process in which, starting from the working structure or concept of a technical system, the design is developed, in accordance with technical and economic criteria, to the point where subsequent design can lead directly to production [2].

Therefore, a crucial objective to be pursued is not only the development of means to support synthesis design tasks but also the analysis of solutions generated upon the intuition and the experience of senior designers, since in modern organizations all the employees should bring a creative contribution to value creation.

The present chapter aims at presenting the authors' experience and vision about the evolution of Computer-Aided systems with respect to synthesis design tasks, with a specific focus on the embodiment of mechanical parts starting from the functional requirements and the related structural and manufacturing constraints.

The next section of the chapter is dedicated to an overview of the related art with the aim of highlighting technological resources as well as the limitations of current systems. The Sect. 2.3 presents the outcomes of a research project (namely PROSIT) aimed at integrating Computer-Aided Innovation systems with PLM tools, while the following proposes the prosecution of the PROSIT project developed by the authors, aimed at embodiment design automation through topological hybridization of partial solutions. The last section proposes a discussion on the expected trends of evolution in this domain and the conclusions of the chapter.

2.2 Related Art

The related art presented here is divided in two subsections, the first dedicated to a brief survey on the role of computers for product development, the second focused on the description of optimization techniques in the field of Computer-Aided Design.

2.2.1 The Role of Computers in the Early Phases of the Product Development Cycle

Computers have gained more and more importance for product development since the dissemination of the first CAD systems prototypes in aerospace industry. Nowadays, they play a crucial role in any industry in detail design tasks, as well as for planning production activities. The so-called PLM (Product Lifecycle Management) systems claim to support any stage of Product Development. In fact, they are extending their domain of application upwards by the preliminary phases of design and by embedding more abstract representations of the product (Fig. 2.1, continuous arrows), but still they are far from systematizing inventive design phases and the link between the development of a conceptual solution and the definition of the product geometry. Indeed, despite it is widely recognized in the relative importance of conceptual design, due to its influential role in determining product's fundamental features, as a matter of fact, CAD/CAE systems are not

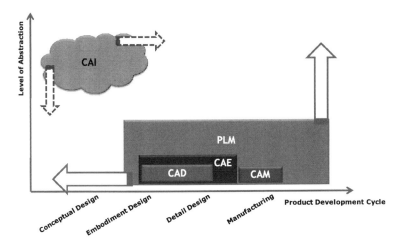

Fig. 2.1 Application domain of computer-based tools within the Product Development cycle

conceived to allow fast input and representation of concept models, and consequently they introduce inertial barriers in experimenting new models of design solutions. Indeed they do not provide any support to designers in developing and expressing their creativity [3, 4].

Recently, Computer-Aided Innovation (CAI) systems have started addressing these lacks [5], trying to leverage the potential of TRIZ [6], the Russian theory for inventive problem solving which constitutes the foundational pillar for most of the CAI software systems. Besides, the domain borders of this emerging technology are still fuzzy and in any case CAI systems suffer of limited interoperability with downstream CAx tools [7].

Thus, a relevant research topic to improve the efficiency of the innovation process is the development of computer-based means for bridging conceptual design with existing PLM systems and the detail design phase, i.e., the extension of the CAI domain toward the embodiment design (Fig. 2.1, dashed arrows).

It is worth to mention that a few preliminary experiments to embed the principles of TRIZ within CAD systems have been attempted with promising, but still not satisfactory, results [8–10]. The main limitation stands in the distance between product models in these two different categories of systems: CAI systems need a more abstract representation, function- or requirement-oriented, with fuzzy topology and/or shape; besides, PLM systems are all structured assuming a more detailed representation of the product, in most cases with explicit geometry and limited possibilities to introduce variations through the control of pre-defined parameters.

A different approach to bridge CAI and PLM systems has been proposed within the PROSIT project [7], whose main outcomes are described in the Sect. 2.3 of this chapter. The logic of PROSIT is to adopt the geometry generation capability of topological optimization systems (briefly overviewed in the next section) as a means to translate the CAI output into a product model manageable by currently available PLM systems. The authors have further developed this concept by developing a semi-automated procedure for conducting the embodiment design phase, through the hybridization of mono-objective optimizations, as described in Sect. 2.4.

Let's consider again the description of the embodiment design phase proposed in [2]: embodiment tasks involve a large number of corrective steps in which analysis and synthesis constantly alternate and complement each other. It is evident that those iterations negatively impact the efficiency of the whole design process, thus a relevant objective for a new method is reducing the need of corrective steps.

According to Kicinger et al. [11] Computer-Aided optimization systems are candidate means to improve design efficiency, thus from this point of view supporting the intuition of the PROSIT project; besides, they claim that topology, shape and size optimization systems can respectively address the needs of conceptual, embodiment and detail structural design. Nevertheless, according to the optimization logic, conflicting requirements are approached looking for the best compromise solution, referred as optimal. Vice versa, it is necessary to highlight that compromise solutions are typically less competitive and have a shorter perspective since, according to TRIZ, technical systems evolve by overcoming, and not compromising, contradictions [6]. It is clear that overcoming contradictions is essential in conceptual design, but avoiding compromise solutions in the embodiment design phase allows to properly exploit valuable concepts.

In conclusions, a straightforward introduction of optimization systems in the product development cycle, even if beneficial for the efficiency of the process, can worsen its effectiveness by pushing the designer to the development of compromise solutions. From this point of view, the development of Computer-Aided systems capable to support the creation of design embodiments beyond the adoption of trivial compromises is a relevant goal for extending the potentialities of CAI systems. According to this statement, the chapter describes an original approach to geometry definition which, despite not involving any inventive act by the designer, is capable to suggest a reduced number of potential topologies and usually results more effective than traditional optimization algorithms.

2.2.2 Design Optimization Systems

Designing by optimization techniques means translating a design task into a mathematical problem with the following basic entities:

- An objective function, i.e., the performance of the system that the designer wants to reach or to improve;
- A set of design variables, i.e., the parameters of the system affecting the objective function;
- Λ set of loading conditions and constraints representing the requirements the system has to satisfy.

The optimization algorithm finds the value of the design variables which minimizes, maximizes or, in general, "improves" the objective function while satisfying the constraints.

The synthesis of product geometry from its functional architecture is an extended perspective for optimization systems; in [9] shape and topological variations of a 3-D model are proposed as a means to generate an optimal geometry through the application of Genetic Algorithms (GAs in the following). Nevertheless, topological and shape variations are obtained through the modification of classical 3-D modeling features, which dramatically limit the design space and impact the practical usability of the proposed method.

The typical classification of optimization systems according to the problems they approach is reported in [11]:

- Topology (layout) optimization, also known as topological optimum design, looks for an optimal material layout of an engineering system;
- Shape optimization seeks optimal contour, or shape, of a structural system whose topology is fixed;
- Sizing optimization searches for optimal cross-sections, or dimensions, of elements of a structural system whose topology and shape are fixed. It is worth to add that a more general definition of this last class of systems refers to parametric optimization, since also other properties of the elements can be assumed as design variables, e.g. the material properties.

Topology Optimization [12] has received extensive attention and experienced considerable progress over the past few years to support design tasks related to the embodiment of functional schemes. It was developed in the structural field but recently it has been applied to address design problems also in other fields such as: fluid dynamics, heat transfer and nonlinear structure behavior: examples of these novel applications of topological optimization can be found in [13, 14].

Topology Optimization determines the optimal material distribution within a given design space, by modifying the apparent material density assumed as design variable. The design domain is subdivided into finite elements and the optimization algorithm alters the material distribution within the design space, according to the Objective and Constraints defined by the designer. The Objective is constituted by one or more system performances that the optimization should improve. Each system performance is quantitatively assessed by an evaluation parameter that is assumed as metric. According to this statement, a mono-goal optimization task tries to improve a single-system performance, while a multi-goal optimization task aims at improving a combination of performances. The constraints of the optimization task represent the operating conditions and the requirements the system has to satisfy. Among them, manufacturing constraints may be set in order to take into account the requirements related to the manufacturing process. Also the regions of the design domain defined as "functional" by the designer are preserved from the optimization process and considered as "frozen" areas by the algorithm. The topology at the end of the optimization process is identified by filtering the resulting material density distribution through a proper threshold having a value included within the interval (0, 1).

Until now, several families of structural topology optimization methods have been developed, a wide literature review is presented in [15].

One of the most established families of topology optimization methods is based on the so-called SIMP approach where SIMP stands for Solid Isotropic Material with Penalized intermediate densities [16, 17]. It uses a gradient-based approach to search the optimal material distribution within the design domain. Thanks to its computational efficiency and conceptual simplicity it has gained a general acceptance in recent years and it is extensively used in the commercial software. The SIMP method is able to deal with optimization problems having a combination of a wide range of design constraints, multiple applied loads and very large 3-D systems. However, as proved by several papers as [12], SIMP gives solutions near to the global optimum only when the optimization problems are convex problems such as those related to the improvement of only one performance of the system (a classical example of an optimization convex problem is represented by the minimization of the compliance of a structure that experiences only one load condition). Unfortunately it is not able to deal with non-convex problems such as multi-objective optimization tasks that are typically related to the improvement of two or more performances of the system. In such cases SIMP could bring to local optimal topologies or converges to an infeasible, i.e., not manufacturable solution. This drawback is common to all the optimization methods based on the mathematical-gradient approach.

Instead of searching for a local optimum, one may want to find the globally best solution in the design domain. For this purpose GAs have become an increasingly popular multi-objective optimization tool for many areas of research. More recently, GAs have been gradually recognized as a powerful and robust stochastic global search method for structural topology optimization [18–20]. Besides, in order to guarantee the robustness of the solution, GAs require more computational resources than the mathematical methods based on the gradient approach. This is due to the high number of design variables that are typically involved in the topology optimization task and this is the main reason why GAs still have not been implemented in commercial CAE tools [21]. Studies on GAs for topology optimization have been performed in recent years, but these attempts are referred to relatively small problems, such as optimization of truss systems with few design variables [22, 23] or 2D problems [24]. Moreover due to the stochastic searching nature of GAs, structural connectivity cannot be guaranteed; this is another main drawback for the application of GAs for topology optimization tasks.

The above literature review shows that the topology optimization techniques based on the mathematical-gradient approach are more efficient than GAs from a computational point of view, but they often bring to local optimum solutions when complex engineering problems have to be solved. Besides GAs present a high robustness in finding global optimal solution for multi-disciplinary problems even if they are not able to deal with a high number of design variables, such as that commonly involved in topology optimization.

However, since the design process has multi-disciplinary characteristics, it implies that improving one performance of a system often may result in degrading another. Such kind of conflicts cannot be solved using topology optimization techniques based on the gradient approach since they are able to focus the design task only to one specific performance to be improved. Besides GAs are design optimization tools that allow to manage multiple goals just by defining complex multi-objective functions but this task requires the definition of a weight to be assigned to each specific goal [25]. Thus, the best compromise solution is generated on the base of an initial assumption made by the designer about the relative importance of the requirements, without taking into account the reciprocal interactions.

2.3 Integrating Computer-Aided Innovation with PLM Systems: The PROSIT Project

As briefly introduced in the previous section, the PROSIT project (www. kaemart.it/prosit), "From Systematic Innovation to Integrated Product Development", aimed at bridging systematic innovation practices and Computer-Aided Innovation (CAI) tools with Product Lifecycle Management (PLM) systems, by means of Design Optimization tools.

The goal of PROSIT was to demonstrate that is possible to define a coherent and integrated approach leveraging on available theories, methods and tools as illustrated in Fig. 2.2.

The rationale behind the adoption of Optimization systems as a bridging means is the following:

- Defining a single multi-objective optimization problem leads to a compromise solution;
- Besides, defining N complementary mono-objective optimization problems, each with specific boundary conditions, leads to N different solutions;
- These solutions can be conflicting and this is the key to find contradictions to be overcome according to the principles of TRIZ.

According to this statement, the PROSIT design flow is structured as depicted in Fig. 2.3. The process starts with the definition of a multi-objective optimization analysis; if the results satisfy the whole set of constraints and requirements the designer can proceed toward the detailed design of the product. Besides, if the output of the multi-goal optimization does not fit the product specifications, a set of single-goal optimization tasks, each representing a specific operating condition and/or a given design requirement, must be defined.

A consequence of the lack of satisfactory solutions to the multi-objective optimization analysis is that the different single-goal optimization tasks lead to conflicting geometries, thus the system must be further investigated in order to extract the "geometrical contradictions" a subclass of TRIZ physical contradiction proposed in [26].

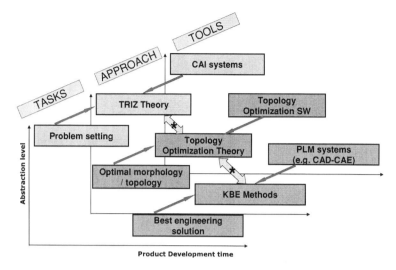

Fig. 2.2 Layered representation of approach/methods and tools supporting the problem-solving tasks of a product-development process

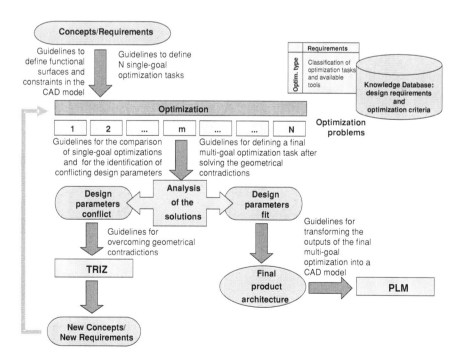

Fig. 2.3 Design flow according to the PROSIT approach

Fig. 2.4 Geometrical
contradiction derived by
the comparison of two
topological optimizations
obtained by applying
alternative boundary
conditions to the technical
system. It is worth to notice
that the density distribution is
not a scalar variable, but a
3-D array representing the
optimized density of each
voxel of the design space

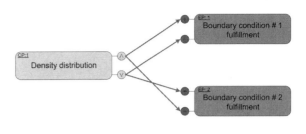

Besides, the analysis and solution of contradictions according to the PROSIT
project was still demanded by the designer, just providing a set of guidelines to
overcome the conflicts according to the specific operational conditions of the
technical system, as described in [7].

2.4 Embodiment Design Automation Through Topological Hybridization of Partial Solutions

Assuming the PROSIT framework as a reference logic, the authors have developed
an original algorithm which approaches the geometrical contradictions emerging
from the comparison of different mono-objective optimizations through a
hybridization process. More in detail, the hybridization is applied to optimized
density distributions obtained by means of traditional topological optimization
systems by assigning just one design requirement for the objective function. Then,
the hybridization, namely DAeMON (hybriDizAtion of Mono-Objective optimizatioNs), is obtained through a TRIZ-inspired manipulation of the topologies to be
combined, as summarized in this section.

The minimal contradiction involves two alternative density distributions arising
from two topological optimizations of the same technical system (TS) where
different boundary conditions are applied, as schematically represented in Fig. 2.4:
the symbols "+" and "−" mean that the behavior of the TS under the ith
Boundary Condition improves and worsens respectively according to the goal
function of the optimization problem. In other words, the diagram in Fig. 2.4
should be read as follows: the density distribution should assume the topology "∧"
in order to improve the behavior of the TS under the Boundary Condition #1, but
then it degrades the behavior under Boundary Condition #2 and should assume the
topology "∨" in order to improve the behavior of the TS under the Boundary
Condition #2, but then it degrades the behavior under Boundary Condition #1.

Such a formulation clearly resembles a classical OTSM-TRIZ contradiction where the density distribution is the parameter under the control of the designer (CP) and the goal function under different Boundary Conditions constitutes the Evaluation Parameters of the Technical Contradiction [27].

More in general, a TS can experience more than two different operating conditions and consequently more than two topologically optimized density distributions can impact the same contradiction. The properties of such a "generalized contradiction" are still under investigation as well as the most effective directions to generate a satisfactory solution [28]. In this chapter only contradictions in the form represented in Fig. 2.4 are taken into account.

2.4.1 Direct Hybridization

Once that a geometrical contradiction in the form represented in Fig. 2.4 has been identified and the mono-objective optimized topologies "∨" and "∧" have been created, the simplest way to perform hybridization is to directly combine the density distributions according to the following formula:

$$\rho(x,\, y,\, z) = \frac{K_1 \rho_1(x,\, y,\, z) + K_2 \rho_2(x,\, y,\, z)}{K_1 + K_2}, \qquad (2.1)$$

where:

- $\rho(x,\, y,\, z)$ is the distribution of density in the design space overcoming the geometrical contradiction;
- $\rho_i(x,\, y,\, z)$ is the distribution of density of the ith mono-goal topological optimization problem;
- K_i is the weight assigned to the result of the ith mono-goal optimization.

The simplest way to assign an appropriate value to the weights K_i is to refer to the potential impact of each loading condition estimated as maximum stress, maximum deformation, strain energy etc. Besides, a more efficient procedure to blend the optimized density distributions $\rho_i(x,\, y,\, z)$ has been proposed in [29] and further developed in [30], by leveraging the potential of Genetic Algorithms to identify the global optimum, but still avoiding the drawbacks GAs meet when applied to classical topological optimization.

In order to clarify the logic of the hybridization process, it is worth considering an exemplary case study taken from optimization literature, such that it is possible to make a comparison between traditional approaches and DAeMON. The problem has been taken from [24] where the multi-objective topology optimization has been performed by means of GAs and concerns the design of a steel plate having an overall dimension of 400 × 300 mm. The plate is discretized with 1,200 (40 × 30) isoparametric plane stress finite elements, assuming an isotropic

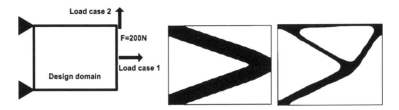

Fig. 2.5 The plate is fully constrained at the corners on the left edge and the forces are alternatively applied on the middle and the upper point of the right edge (*left*). Topologies obtained through mono-objective optimization under load case 1 and density threshold = 0.27 (*middle*) and under load case 2 and density threshold = 0.83 (*right*)

material with Young's modulus equal to 210 GPa and Poisson's coefficient equal to 0.3.

As depicted in Fig. 2.5 left, the plate undergoes two different point loads, a vertical force and a horizontal load, both with a magnitude of 200 N. By performing two different mono-objective optimizations according to the PROSIT logic, two conflicting geometries emerge: Fig. 2.5 shows the topologies "∨" (middle) and "∧" (right) in analogy with the model of geometrical contradiction represented in Fig. 2.4. The direct hybridization brings to a set of solutions which depend on their blending proportion (Fig. 2.6).

The resulting topology appears more effective (in terms of stiffness related to the overall mass) than a standard multi-objective optimization. At the same time, a comparison with the results obtained with a GA topological optimization presented in [24] reveals similar mechanical performances, but computational efforts an order of magnitude smaller. Further details on this comparison and to other similar case studies are reported in [29].

2.4.2 Rotations and Translations as Possible Mutations for Extended Hybridization

Indeed, the investigation carried out by the authors about many different geometrical contradictions and related solutions [26] revealed that direct hybridization is not the only solution strategy which can be adopted to overcome a contradiction. Further relevant solution paths can be associated to TRIZ heuristics [6]:

- Different orientation of a geometrical feature, i.e., a rotation of a geometrical element, or in TRIZ terms, "Another Dimension" (Inventive Principle #17);
- Multiple copies obtained by a translation of a geometrical feature, as suggested from the trend of evolution Mono-Bi-Poly of homogeneous systems (Inventive Standard 3.1.1) applied to geometrical features (Fig. 2.7, above);
- A combination of the above, i.e., the trend Mono-Bi-Poly applied to systems with shifted characteristics (Inventive Standard 3.1.3) obtained by introducing

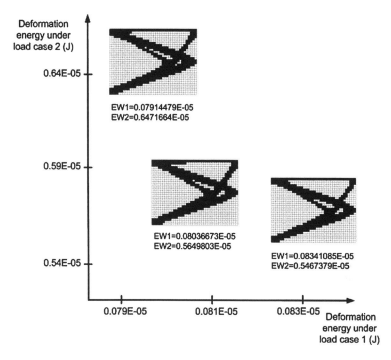

Fig. 2.6 Hybrid solutions obtained through a different blending of the mono-objective optimizations related to the two different operating conditions

Fig. 2.7 Mono-Bi-Poly transformation applied to geometrical features (*above*). Exemplary bi-features obtained by a combination of rotations and translations of the original geometry (*below*)

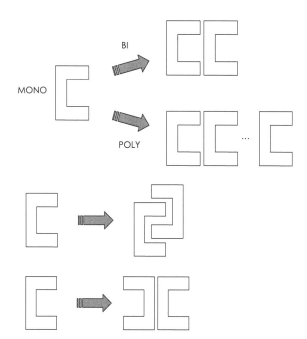

multiple copies of a geometrical feature, each with a proper position and orientation; the simplest case is obtained by duplicating a geometrical feature by means of a mirror operation (Fig. 2.7, below).

A general expression capable to represent all the above solution strategies is the following:

$$\rho(x, y, z) = \frac{\sum_{i=1}^{N} \sum_{j=1}^{Mi} K_{ij}\left([ROT](x, y, z)^T + (x_0, y_0, z_0)_{ij}^T\right)}{\sum_{i=1}^{N} \sum_{j=1}^{Mi} K_{ij}}, \qquad (2.2)$$

where

- N is the overall number of conflicting mono-goal optimizations (two if a classical TRIZ contradiction model is adopted, as represented in Fig. 2.4);
- M_i is the number of "copies" of the i-th solution (step of a mono-bi-poly trend);
- K_{ij} is the weight assigned to the jth copy of the i-th distribution of density;
- $[ROT]_{ij}$ is a general expression coming from Linear Algebra, describing a matrix 3 X 3 that contains Direction Cosines related to the angles between the coordinate axes of the initial and the rotated system, thus, in such context, this term represents the rotation applied to the j-th copy of the i-th distribution of density;
- $(x_0, y_0, z_0)_{ij}$ is the translation applied to the j-th copy of the i-th distribution of density.

The appropriate values for M_i, K_{ij}, $[ROT]_{ij}$ and $(x_0, y_0, z_0)_{ij}$ are still under investigation; nevertheless, a typical combination for axial-symmetric density distribution is: $M_i = 1$; $K_{ij} = 1$; $[ROT]_{11} =$ identity matrix (no rotations); $[ROT]_{21}$ is a rotation around the axis of the system, the angle being calculated as half the periodicity of the geometrical feature; $(x_0, y_0, z_0)_{i1} =$ null vector (no translations).

An exemplary application of hybridization of rotated topologies refers to the re-design of a motor-scooter wheel [31]. The test case has been inspired by a real case study developed during a collaboration of the authors with the Italian motorbike producer Piaggio. The goal of the project was the design of a plastic wheel for light moto-scooters mainly aimed at costs reduction, of course without compromising safety and mechanical performances.

The traditional approach used in Piaggio to assess the conformity of a wheel to requirements consists in three different experimental tests:

1) Deformation energy under high-radial loads/displacements (simulating an impact against an obstacle);
2) Fatigue strength under rotary bending loads (simulating the operating conditions such as curves);
3) Fatigue strength under alternate torsional loads (simulating the accelerations and decelerations).

Fig. 2.8 Output topologies obtained by topological optimizations: boundary conditions (loads and constraints), optimization constraint (overall mass), optimization objective and density threshold are the same for all four instances. Only the number of the pattern repetition is clearly different

These tests have been adopted as reference criteria for topology design optimization, under the constraint of manufacturability through injection and molding and the goals of minimizing mass and maximizing the stiffness distribution on the rim wheel. The optimization problem has been set up as follows:

- Objective Function: maximize wheel stiffness;
- Constraints: several upper limits for the mass of the wheel; manufacturing constraints for injection molding process;
- Loading conditions: radial and tangential loads applied on the rim of the wheel.

Rim profile and hub have been defined as non-design areas since they are functional surfaces. The optimization task led to several topologies having different number of spokes (Fig. 2.8). Their compliance to the design criteria described above has been checked through virtual simulations.

Results revealed that three and six spokes wheels widely satisfy the deformation energy test only when the radial load is applied on the areas of the rim directly supported by a spoke while, when the radial load is applied among them, the proof fails. The other topologies never satisfied the deformation energy criterion while all meet fatigue strength requirements (2, 3).

A deeper investigation of the radial stiffness distribution along the wheel rim has been performed for each optimized geometry (Fig. 2.9). As supported also by intuition, when the number of spokes rises, the stiffness of the rim on spokes decreases, while it increases among the spokes.

According to these results a contradiction appears: a smaller number of spokes provides the highest radial stiffness in the areas of the rim directly supported by the spokes, but the deformation between two spokes is maximum. A bigger number of spokes allows to obtain a more uniform stiffness distribution along the rim but with low overall values. This technical contradiction can be modeled as shown in Fig. 2.10.

Fig. 2.9 Normalized radial stiffness distribution evaluated on the wheel rim for different topologies: radial force applied on the spokes (*dark*) and in the middle between two adjacent spokes (*light*)

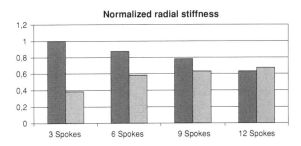

Fig. 2.10 Model of the technical contradiction: EP1 is the stiffness on spokes, EP2 is the stiffness among spokes

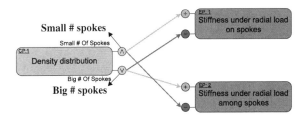

Taking into account these considerations, "three spokes" and "nine spokes" geometries have been selected to produce an improved "manipulated" topology through Formula (2.2). The goal is the definition of a new topology, not identified by standard optimization systems, with a higher mechanical performance.

Taking into account the functional surfaces, the hub axis is assumed as reference to apply the transformation. The rotation is defined as a half of the angular periodicity of the nine spokes wheel, thus 20°: such a value provides the minimum overlap between the original distributions of density. Figure 2.11 shows the profile of the original distribution of densities (3 and 9 spokes wheels) and the result of the manipulation; as a result of the density combination, a "Y" shaped spoke is suggested. It is worth to notice that such a topology is definitely different from any result provided by the optimization systems.

A preliminary design of a Y-shape spoke wheel has been developed in order to compare its radial stiffness with the mechanical performance of the original geometries. Figure 2.12 summarizes the results of such a comparison.

The analysis revealed that the suggested topology is 20% lighter than both the "three spokes" and "nine spokes" configurations. The "Y" version gives also an improvement of the rim radial stiffness among spokes.

Even if the stiffness evaluated on spokes worsens with respect to the "three spokes" wheel, "Y" configuration satisfies the deformation energy design criterion.

Fig. 2.11 *Above*: conflicting distributions of density according to the contradiction modeled in Fig. 2.10. (same overall mass). *Below*: density distribution automatically obtained by the application of Formula (2.2) to the conflicting pair (*left*) and exemplary interpreted geometry (*right*). The darkness of the images is directly proportional to the optimized density

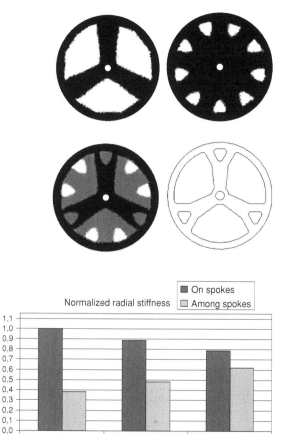

Fig. 2.12 Comparison of radial stiffness distribution among "three spokes", "Y" and "nine spokes" wheels. "Y" has an improved stiffness among spokes with respect "three spokes". The behavior is similar to the "nine spokes" wheel but with an improved stiffness on spoke. "Y" is 20% lighter than the other configurations

2.5 Trends of Computer-Aided Design: Discussion and Conclusions

Among the trends which characterize the evolution of Computer-Aided tools supporting product development, a relevant aspect is the extension of the domain of application described in Sect. 2.2 and depicted in Fig. 2.1 Indeed, the engagement of this trend can be seen as a transition from task-oriented applications to process-oriented systems: the former CAE tools were able to speed-up and sometimes automate several engineering tasks, but the integration was limited to product data exchange formats. Then, PLM systems emerged as a "strategic business approach that applies a consistent set of business solutions in support of the collaborative creation, management, dissemination, and use of product definition information across the extended enterprise from concept to end of life–integrating people, processes, business systems and information" [32].

As observed above, current PLM systems are effectively integrated just with CAD-CAE applications, but their efficiency is still poor for the preliminary design phases. In order to extend the domain of application of PLM system, a necessary transition is to introduce a modeling approach capable of representing a product at different detail levels, from the functional requirements of the earliest stages of conceptual design, to the constructive details of the manufacturing stage.

The present chapter summarizes the contribution of the authors toward the definition of a model capable to represent a mechanical part since the intermediate stage of embodiment design. In fact, the density distributions generated by topological optimizations of mono-objective tasks can be seen as elementary customized feature for the definition of the geometry of a certain mechanical part during the embodiment stage, when its functional role must be translated into a geometry to be manufactured and coupled with other subsystems.

It is worth to highlight some characteristics of these customized modeling features:

- As mentioned in Sect. 2.2, the result of a topological optimization is a distribution of density so that each cell of the design space assumes a fuzzy value between 0 and 1, which in turn means that boundaries are not rigid as it happens also with classical free-form modeling features; in fact, a density distribution can produce both topological and shape variations while, apart few exceptions, parametric modifications of a free-form surface produce just shape variations;
- Compared with free-form surfaces where a shape variation is obtained by moving many control nodes, the output of a topological optimization produces different specific geometries by editing just one parameter, i.e., the threshold value of the density discriminating between void and filled space;
- These customized modeling features can be combined according to a general Formula (2.2) which embeds several TRIZ inventive principles, thus inheriting the potential to overcome the emerging geometrical contradictions, as exemplarily shown in Sect. 2.4;
- It is important to note that the proposed hybridization approach usually leads to very different topologies with respect to traditional design multi-objective optimization; the resulting geometry has often better performance than the traditional one;
- The specific hybridization strategy can be governed by means of GAs, thus inheriting their capability to search for a global optimum, but at the same time avoiding the unaffordable computational demand of the application of GAs to topological optimization;
- Compared with GAs, the DAeMON approach is also intrinsically robust against the definition of non-manifold and/or non-manufacturable geometries.

The adoption of topological optimization as a bridging element between the generation of a concept and the development of a detailed solution has also further advantages.

In fact, topological optimization just requires to define the design space and the functional surfaces of the part to be designed. These surfaces, can therefore be linked to the function delivered by the part itself, thus creating a connection between the abstract product representation of the conceptual design stage with the geometrical details of the following phases. Moreover, this vision fits also with the trend toward the knowledge integration into CAD systems already approached by Knowledge-Based Engineering (KBE) systems to automate configuration tasks of modular products [33].

With the aim of further extending the applicability of the DAeMON logic to more complex design problems, as well to fuzzier problem situations which emerge in the development of innovative projects, the authors are also working to define a systematic link between the outcomes of a conceptual analysis made with OTSM-TRIZ techniques [27] and the set up of the mono-objective optimizations which constitute the starting point of the proposed methodology [34].

The long-term vision is a transition to a new generation of CAD systems which will guide the designer through a systematic analysis of the task to be accomplished until the functional architecture of the system has been defined [35]. Then, each functional component should be embodied starting from the definition of the optimal topology by means of the hybridization of the customized modeling features, i.e., density distributions, described in this chapter. According to this perspective, a further relevant direction of investigation is the definition of a link between a density-based representation of geometry and the classical feature-based approach, still more convenient for the following manufacturing stages.

Acknowledgments The author would like to thank Alessandro Cardillo from Politecnico di Milano and Francesco Saverio Frillici of University of Florence for their contribution to the development of this research.

References

1. Cavallucci D, Eltzer T (2007) Parameter network as a means for driving problem solving process. Int J Comput Appl Technol 30(1/2):125–136
2. Pahl G, Beitz W (2007) Engineering design: a systematic approach, 3rd edn. Springer-Verlag, London
3. Qin SF, Harrison R, West AA, Jordanov IN, Wright DK (2003) A framework of web-based conceptual design. Comput Ind 50(2):153–164
4. Tovey M, Owen J (2000) Sketching and direct CAD modeling in Automotive Design. Des Stud 21(6):569–588
5. Leon N (2009) The future of computer-aided innovation. Comput Ind 60(8):539–550
6. Altshuller GS (1979) Creativity as an exact science. Gordon and Breach Science Publishers, New York
7. Cugini U, Cascini G, Muzzupappa M, Nigrelli V (2009) Integrated computer-aided innovation: the PROSIT approach. Comput Ind 60(8):629–641
8. Chang HT, Chen JL (2004) The conflict-problem-solving CAD software integrating TRIZ into eco-innovation. Adv Eng Software 35:553–566
9. Leon N, Cueva JM, Silva D, Gutierrez J (2007) Automatic shape and topology variations in 3D CAD environments for genetic optimization. Int J Comput Appl Technol 30(1/2):59–68

10. Albers A, Leon N, Aguayo H, Maier T (2007) Comparison of strategies for the optimization/innovation of crankshaft balance Trends in computer-aided innovation. Springer, New York, pp 201–210
11. Kicinger R, Arciszewski T, De Jong K (2005) Evolutionary computation and structural design: a survey of the state-of-the-art. Comput Struct 83:1943–1978
12. Rozvany GIN (2001) Aims, scope, methods, history and unified terminology of computer-aided topology optimization in structural mechanics. Structu Multidiscip Optim 21(2):90–108
13. Hutabarat W, Parks GT, Jarret JP, Dawes WN, Clarkson PJ (2008) Aerodynamic topology optimisation using an implicit representation and a multi objective genetic algorithm. Artificial evolution, Lecture. notes in computer science, vol 4926. Springer, Berlin/Heidelberg, pp 148–159
14. Bruns TE, Tortorelli DA (2001) Topology optimization of non-linear elastic structures and compliant mechanisms. Comput Meth Appl Mech Eng 190(26/27):3443–3459
15. Rozvany GIN (2009) A critical review of established methods of structural topology optimization. Structu Multidiscip Optim 37:217–237
16. Bendsoe MP, Kikuchi N (1988) Generating optimal topologies in structural design using a homogenization method. Comput Meth Appl Mech Eng 71:197–224
17. Bendsoe MP, Sigmund O (2003) Topology optimization theory methods and applications. Springer Verlag, Berlin Heidelberg
18. Wang SY, Tai K (2005) Structural topology design optimization using genetic algorithms with a bit-array representation. Comput Meth Appl Mech Eng 194(36–38):3749–3770
19. Wang SY, Tai K (2004) Graph representation for evolutionary structural topology optimization. Comput Struct 82(20,21):1609–1622
20. Neumaier A (2004) Complete search in continuous global optimization and constraint satisfaction. Acta Numerica 13:271–369
21. Zuo ZH, Xie YM, Huang X (2009) Combining genetic algorithms with BESO for topology optimization. Structu Multidiscip Optim 38:511–523
22. Ryoo J, Hajela P (2004) Handling variable string lengths in GA-based structural topology optimization. Structu Multidiscip Optim 26:318–325
23. Wang SY, Tai K (2005) Structural topology design optimization using genetic algorithms with a bit-array representation. Comput Meth Appl Mech Eng 194:3749–3770
24. Aguilar Madeira JF, Rodrigues H, Pina H (2005) Multi-Objective optimization of structures topology by genetic algorithms. Adv Eng Software 36:21–28
25. Spath D, Neithardt W, Bangert C (2002) Optimized design with topology and shape optimization. P I Mech Eng B-J Eng 216(8):1187–1191
26. Cascini G, Rissone P, Rotini F (2007) From design optimization systems to geometrical contradictions. Proceedings of the 7th ETRIA TRIZ Future Conference, Kassel University Press, Kassel pp 23–30
27. Cavallucci D, Khomenko N (2007) From TRIZ to OTSM-TRIZ: addressing complexity challenges in inventive design. Int J Prod Dev 4(1–2):4–21
28. Eltzer T, De Guio R (2007) Constraint based modelling as a mean to link dialectical thinking and corporate data. Application to the Design of Experiments. Trends in computer-aided innovation, Springer, New York, pp 145–156
29. Cardillo A, Cascini G, Frillici FS, Rotini F (2011) Computer-Aided embodiment design through the hybridization of mono objective optimizations for efficient innovation process. Comput Ind 62(4):384–397
30. Cardillo A, Cascini G, Frillici FS, Rotini F (2011) Multi-objective topology optimization through GA-based hybridization of partial solutions. Submitted for publication to Engineering with Computers, 2011
31. Cascini G, Rissone P (2004) Plastics design: integrating TRIZ creativity and semantic knowledge portals. J Eng Des 15(4):405–424
32. CIMdata (2002) PLM Market Analysis Report. http://www.cimdata.com, Accessed 24 February 2011

33. Cascini G (2004) State-of-the-Art and trends of Computer-Aided Innovation tools–toward the integration within the Product Development Cycle. Building the information society, vol 156. Kluwer Academic Publishers, Toulouse, pp 461–470
34. Cardillo A, Cascini G, Frillici F, Rotini F (2011) Embodiment design through the integration of OTSM-TRIZ situation analysis with topological hybridization of partial solutions. Proceedings of the 18th international conference on engineering design (ICED11), Copenhagen
35. Becattini N, Borgianni Y, Cascini G, Rotini F (2011) Model and algorithm for computer-aided inventive problem solving. Computer-Aided Design, In Press, Available online 26 February 2011, doi:10.1016/j.cad.2011.02.013

Chapter 3
Methods and Tools for Knowledge Sharing in Product Development

Marco Bertoni, Christian Johansson and Tobias C. Larsson

Abstract The emerging industrial business partnerships, which feature cross-functional and cross-company development efforts, raise the barrier for the establishment of effective knowledge sharing practises in the larger organization. This chapter aims to highlight the role of knowledge as a key enabler for effective engineering activities in the light of such emerging enterprise collaboration models. Knowledge enabled engineering (KEE) is presented as an approach to enhance the extended organization's capability to establish effective collaboration among its parts, in spite of different organizational structures, technologies or processes. KEE is analyzed in its constituent parts, highlighting areas, methods and tools that are particularly interesting for leveraging companies' knowledge sharing capabilities.

3.1 Introduction

The development process, as all work processes, involves some combination of four types of "work" (Fig. 3.1), which can be thought about as a continuum [1]. At one end of this continuum we find the entirely independent effort of an individual to produce some kind of product or service. Next, we find dependent work, where one applies some level of effort to make use of someone else's product or service. Along this continuum, participants need to share information across disciplines,

M. Bertoni (✉) · C. Johansson
Luleå University of Technology, SE-97 187 Luleå, Sweden
e-mail: marco.bertoni@ltu.se

C. Johansson
e-mail: christian.johansson@ltu.se

T. C. Larsson
Blekinge Institute of Technology, SE-371 79 Karlskrona, Sweden
e-mail: tobias.larsson@bth.se

Fig. 3.1 The four levels of
interaction

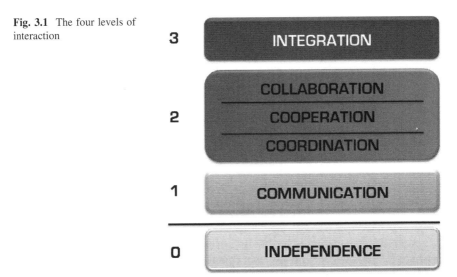

and other barriers, to achieve a mutual objective. Eventually we find the full
integration of knowledge acquired through specialists in a range of disciplines,
which enables a team to optimize the work to be performed.

Collaboration is a term mainly referring to the third stage of the above continuum,
i.e. to "people working together on common tasks, or to the aid extended to a thing or
a person to produce or create something new" [2], just as a blacksmith and a
glassblower can collaborate on a piece of art for mutual benefit. Fundamentally, it is
based on the idea that working together will allow collaborating partners to create
something superior to what any one entity could have created alone.

This concept is of high interest both in the business-to-business and business-to-
consumer industry [2]. In the latter, customer service manufacturers may try to
develop their collaborative skills to better reach out to their customers and to
maintain constant contact with them. On the other hand, collaborative product
development processes, involving people, processes and technologies across
multiple organizations working in the same line of business, is becoming the
industry standard in light of globalization and outsourcing. True collaboration is
still a pipe dream for large product development projects. The difficulties recently
encountered by, for example, the aeronautical industry [3, 4] clearly highlights that
effective collaborative development processes require a quantum jump in the way
information and knowledge is shared in the extended organization.

3.2 Motivation and Objectives

The ultimate goal of any enterprise collaboration means is to support a more
effective and trustworthy decision-making. In essence, "it's all about getting the
right information to the right people at the right time so they can do their jobs more

effectively" [5]. To enable conscious decisions, enhanced knowledge sharing methods and tools are needed to make sure that all design stakeholders contribute with their experience and skills in such decision-making processes, while protecting intellectual ownership.

The main objective of this chapter is to highlight the role of knowledge as a key enabler for effective engineering activities, in the light of the emerging enterprise collaboration trends.

The chapter initially describes the different levels of interaction that are featured in modern business partnerships, highlighting the difficulties in establishing effective collaboration, and consequently knowledge sharing practises, when dealing with cross-functional and cross-company development efforts.

The chapter further introduces the concept of KEE, an acronym that summarizes the intent of providing the extended organization with the capabilities to establish effective knowledge collaboration among its parts, in spite of different organizational structures, technologies or processes. KEE is eventually analyzed in its constituent parts, highlighting areas, methods and tools that are particularly interesting for leveraging companies' knowledge sharing capabilities.

3.3 Four Levels of Interaction

Product development processes can be differentiated on the basis of the interaction required for addressing the common development target. Due to cost-benefit considerations the intensity and design of integration have to be based on the requirements of the existing development situation [6]. It is possible to outline four major levels of interaction between organizations and teams (Fig. 3.1):

1. Level 0: Independence means that companies do not interact in any form during the development tasks and do not share any kind of data or information during the process.
2. Level 1: Communication refers to the process of transferring information from one source to another, and is at the basis of any collaborative task. It simply refers to the act of exchanging information by the use of some kind of media and does not imply working toward a common goal.
3. Level 2: Coordination, Cooperation, and Collaboration refers to the association of a number of persons for their common benefit and to collective action in pursuit of a common goal. Autonomous, independent, federated and loosely coupled systems can exchange services while continuing their own logic of operation, similarly to Virtual Enterprise [7] agreements.
4. Level 3: Integration refers to the concepts of coherence and standardization indicating a 'tightly coupled' system where components are interdependent and cannot be separated. This level points to the more traditional single-OEM structure; a homogeneous organization, composed by interdependent parts and standard languages, methods and tools.

It is possible to define three levels of collaboration intensity for product development: coordination, cooperation and collaboration [6]. Which type should be chosen depends on the required interaction intensity caused by the interaction need of a given development situation, and typically increases with tighter process-related interdependencies between the subtasks, and with the necessity of integrating the knowledge of the development partners.

- Coordination is the act of managing interdependencies between activities [8] and the support of interdependencies among actors [9], regulating diverse elements into an integrated and harmonious operation to accomplish a collective set of tasks.
- Cooperation, similarly, refers to the association of a number of persons in the pursuit of common goals [10], able to adhere to some kind of dialogue rules [11], i.e. setting out maxims (of quantity, quality, relevance and manner) to which it can be assumed parties in a dialogue are adhering on an utterance-by-utterance basis [12].
- Collaboration is a special case of cooperation and implies the joint execution of one task where the individual, private aims of the participants are more closely aligned with each other. People need to work together to reach the desired outcome rather than that outcome being achieved through 'selfish' participation constrained by contextual factors [11]. Collaboration may be seen as the highest degree of interdependency between autonomous and independent parties.

3.4 Emerging Enterprise Collaboration Models

In those areas where products are complex and require large investments it is common for companies to collaborate in joint ventures where risks, costs, revenues and profits are shared among the partners. In such a context, two different types of partnerships are of particular interest with a focus on collaboration: the Extended Enterprise (EE) [13] and the Virtual Enterprise (VE) [7].

The EE is a form of joint venture where one company, usually named Original Equipment Manufacturer (OEM) is in charge of integrating the parts of a product, produced and designed by satellite companies, into the whole [14]. The objective with forming an EE is building long-term relationships across the value chain, both to share risks, knowledge and resources [15] and to reduce costs and time-to-market [16].

The VE spans over multiple EEs and involves several different OEMs, which are asked to temporarily collaborate to design and manufacture new products. The VE partners share costs, skills and core competencies to deliver solutions that none of the individual partner could have done on their own [17]. Figure 3.2 visualizes the relationship between the EE and the VE.

While the EE focuses more on long-term partnerships, the VE is a very short-term and flexible endeavor [19], where companies participate for a shorter time

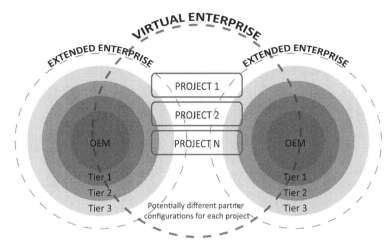

Fig. 3.2 Virtual enterprise vs. extended enterprise [18]

span to satisfy a business opportunity relatively quickly [7, 13]. The cross-organizational nature of the VE allows companies to be flexible with the use of resources [13], moving people in and out of the individual organisations as the conditions changes and different competencies are in demand.

3.5 Barriers and Inhibitors for Effective Knowledge Sharing

Knowledge sharing is a key aspect to any strategic alliance, since fundamentally, companies enter these partnerships to tap into the knowledge bases of others to expand the offerings they can make. Managing such knowledge is one of the most crucial aspects for the successful implementation of a virtual working paradigm. However, a number of collaboration barriers, which are often not visible in a co-located and "physical" organization, inhibit the EE and VE effectiveness to share knowledge across teams, functions and companies. Considering Allen's [20] observations that people are likely to seldom collaborate if they are more than 50 feet apart, the challenges of globally distributed organizations are immense.

As Brown and Duguid [21] point out, "knowledge usually entails a knower" (p. 119), i.e. knowledge sharing relies on humans taking on the role of creator, carrier, conveyor and user. This distinguishes knowledge from information sharing, which can usually happen "outside" humans and without the direct influence of human beings. Information is normally treated as independent and more-or-less self-sufficient, whereas knowledge is usually associated with someone (e.g. "where is that information" vs. "who knows that?"). In a virtual organization the right people to ask (knowledge owners, hidden experts, lead users, etc.) are even more difficult to locate than in a traditional (co-located) engineering situation [18].

Fig. 3.3 Virtual distance
[18]

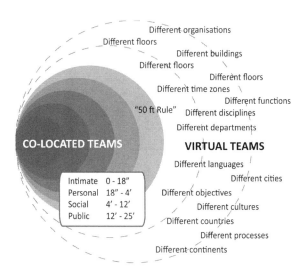

Furthermore, work practises may differ substantially between the different parts of the consortium, which generates strong needs to share knowledge-of-practises and be able to locate people with credibility [18]. In co-located teams, knowledge-of-expertise is usually built-up naturally and quickly, through the interactions with others. People gradually learn about the strengths and weaknesses of their colleagues, and when faced with a problem they often have a very good idea of whom to approach for help.

Also, engineers must be able to trust the information they receive—so called trust-in-expertise [18]. This trust fundamentally boils down to a trust-in capacities and abilities, and people need to make continuous assessments about whether or not they can rely on recommendations or advices that their colleagues offer. Another common problem of cross-site collaboration is the delay in resolving work issues [22], due to the difficulties of finding the right person to talk to, initiating contact, and discussing possible ways to deal with the issue. As physical distance increases, tacit knowledge (e.g. intuition, individual perception and rules of thumb) is not spread as easily through informal communication channels, and must therefore be made clearer to be used throughout the company. Figure 3.3, adapted from Lipnack and Stamps' [23], illustrates some of the barriers related to "virtual distance".

From a company perspective, a main issue is finding the best trade-off between sharing and shielding core knowledge assets. The "coopetitive" [24] nature of the VE arrangements makes knowledge sharing problematic because knowledge used for cooperation may also be used for competition in other projects [25]. Companies are feel hesitant to share knowledge if they feel that what they gain is less than what they give away [26], so the problem is to understand what has to be shared, and in what form, to sustain the project without losing the main source of competitive advantage.

Last, but not least, the loss of face-to-face interaction have an inhibiting effect on morale, loyalty and performances of employees [27], who may "interact" but not collaborate. If on one side technology can cross boundaries, people often can not. Plenty of social and behavioral issues have yet to be solved to implement effectively the virtual working paradigm across organizations.

3.6 KEE: Enabling Engineering Activities Through Knowledge

In an ideal world, companies should be able to use their collective knowledge to make better and faster decisions, reducing lead-time and improving robustness of strategic engineering activities. Engineers should concentrate on the more intellectual parts of engineering work, rather than spending time doing dull and cumbersome routine work.

The term Knowledge Enabled Engineering (KEE) [28] collects a wide range of methods and tools that intend to enable such ideal scenario, supporting a knowledge sharing activity that spans long product life cycles within multi-disciplinary, multi-company and multi-cultural environments, across supply chain relationships [29]. The main purpose of KEE is to make as much knowledge as possible available early in the product development process, to allow more design iterations via simulation [30], reducing the lead-time for tedious analysis model generation and thus allowing more design concepts to be studied [31].

KEE originates from the work done in the area of knowledge engineering (KE) and knowledge based engineering (KBE). However, while KBE is often associated with commercial systems providing demand driven, object oriented programing languages for rule-based execution, KEE is intended as a more generic and methodological-oriented term [32] with emphasis on context and rationale. KEE systems are intended to support three main tasks: (1) promoting the use of fitting-for-purpose models to support the capturing of engineering knowledge in different engineering activities, (2) supporting an iterative process where the capturing of engineering knowledge and the automation of the engineering activities is done simultaneously and (3) capturing the underlying process of generating and evaluating solutions concepts in a computerized system to ensure repeatability and improve the quality of the output.

The KEE research aims to develop, validate and exploit methods to identify, model, store, retrieve, reuse and share product/service relevant knowledge spanning long product life cycles. The following sections aim to spotlight the most crucial areas to leverage the company's capability to enable engineering activities through a better use of their knowledge assets.

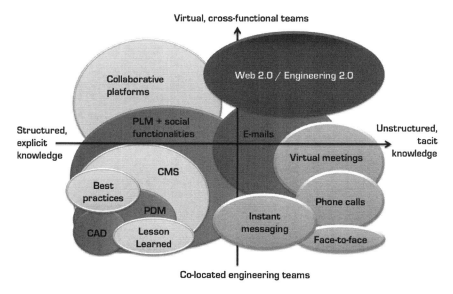

Fig. 3.4 Engineering 2.0 positioning

3.6.1 Enabling Informal Knowledge Sharing Between Dislocated Design Teams

The advent of Web 2.0 technologies has brought a new culture of sharing information on the Web where users can actively create, store, edit, access, share and distribute the content to larger audiences. This 'bottom-up' sharing mode is opposed to the predefined, 'top-down' structure of most of the existing knowledge management tools, pushing each individual to maintain their own space for which they have complete control over the information they like to share.

Moving from McAfee's [33] concept of Enterprise 2.0, Larsson et al. have coined the term "Engineering 2.0" [34], which borrows the more general Web 2.0 concept and translates it into a cross-functional engineering context. Engineering 2.0 promotes the use of lightweight (compared with traditional CAD/PDM/PLM) and bottom-up tools to support informal knowledge sharing across functions and companies in a VE setting (Fig. 3.4).

Lightweight because the purpose is to develop and implement solutions that require little time and effort to setup, use and maintain. Bottom-up because they do not impose a pre-defined structure, but rather let structures evolve over time as an almost organic response to the activities, practises and interests of the knowledge workers. Weblogs, wikis, social networks, RSS feeds, tagging, microblogs, instant messaging, discussion forums, social bookmarking and mash-ups are few examples of Web 2.0 technologies that can be successfully adopted in an engineering setting.

A recent McKinsey survey [35] has shown that 2/3 of 1,700 companies interviewed worldwide have investigated or deployed Web 2.0 tools to support their product development activities. However, their use is limited today to a comparably small part of the entire development process, mainly with the intent of gathering ideas and feedback on the product from the customer base [36].

Web 2.0 tools are used mainly to enable effective communication within dislocated teams, particularly in the initial stages of a software development project [37]. The Microsoft Quest internal communications system [38], the wiki-like environment proposed by Ciavola et al. [39] and IBM Dogear [40] are a few examples of social software supporting learning, sharing and collaboration between researchers and professionals in the design activity.

Lightweight approaches have been also proposed to support product development projects' documentation [41], e.g. using wiki-style collaboration tools to create assessment reports [42] or to maintain rule-based systems as they grow [36]. Furthermore, several groups are proposing ways to complement CAD/PDM/PLM tools with social functionalities, leveraging social interaction and collaborative features, among global design teams, Vuuch [43], for instance, is a plug-in for Pro/ENGINEER or Dassault Systemes' SolidWork, to initiate, monitors and manages design discussions directly from the CAD environment to organize design discussions by associating them to the product Bill of Material (BOM).

3.6.2 Enabling Fit-for-Purpose Reuse of Past Design Knowledge

The key element to successful knowledge reuse is to understand a designer's reuse intention [44]. The designer's information seeking behavior, in fact, depends on task-dependent procedures [44]. It means that the usual search queries alone are not sufficient to address the designers' information needs. Rather, if the context for the query is known, it is possible to anticipate the type of result that will be useful, and refine the query accordingly, providing more tailored knowledge to people with similar profiles [45]. Context-aware applications may help to increase the relevance of the knowledge they retrieve to support a given task.

Context-awareness can be defined as "the ability of a computing device to detect and sense, interpret and respond to aspects of a user's local environment and the computing devices themselves" [46]. Context-aware system have gained popularity in the 1990s to address problems raised by the usage of mobile terminals [47], with the purpose of (1) retrieving information and execute commands for the user manually based on available context, (2) automatically executing a service based on the current and available context, (3) tagging the context to information for later retrieval. Context may refer, therefore, both to a physical dimension (time, space, device) and internal/logical dimension, i.e. the user's goals, tasks, work context, business processes, the user's emotional state [48].

One of the most notable examples of Context-aware applications supporting collaborative product development is the knowledge-enabled solution platform

(KESP) [49] developed within the EU FP6 VIVACE project [28]. The KESP is an intelligent knowledge assistant that automatically provides the engineer with the contextualized knowledge elements (K-El) that s/he needs during his/her daily-design activity. The KESP is both a multi-source context-based application and a self-learning software system that enables the user to perform multi-source searches for all the K-El applicable to his or her engineering context. It describes the engineers' profile by using six contextual dimensions (Product, Process, Project, Gate, Role, Discipline) and enables to manage links between knowledge and contexts of applicability, relying on similarities between contexts to provide the knowledge workers K-El relevant for their task. The driving idea behind the KESP is the possibility to associate an element with the most suitable context in which it should be applied. For some K-El, such as norms, it may be quite straightforward to define beforehand the domain of applicability. For others it may be more complex. To overcome this obstacle, the KESP provides a way to learn the relationship between a K-Element and the description of the context in which it has been applied, so making easier for engineers to understand what documents/information is more relevant for their job.

3.6.3 Enabling Effortless Management of Design Rationale

A recent study [50], presenting the results from a UK survey about requirements of managers and engineers in design and service, has shown that "rationale" (34.9%) is the most mentioned category in terms knowledge and information needs in engineering. Understanding the reasons why a system has been designed in a certain way, or the information about what design options have been considered but rejected, is necessary to understand, recreate or modify a design, although rarely adequately captured in a systematic and usable format, [51], but rather scattered throughout a collection of paper documents, notebook entries and the memory of the designers [52].

Design intent capture systems aim to capture and store information about past design solutions and their rationale in a repository that is independent of human memory [53]. They can be divided into solutions aiming to "communicate" or "document" design knowledge [51]. The first category [54] aims to capture all communications among team members during design meetings, but do not encourage the synthesis and retrieval of high-level design decisions (assumptions, constraints, philosophies) [51]. The second category documents design decisions to enable people outside the project group to understand, supervise and regulate what is done by the team [55], although not usually capturing why a particular design was not chosen [51]. Furthermore, automatic approaches aim to capture the communication among the designers and design teams, or between the designer and the design support system, without asking user intervention [56]. Manual tools, instead, require the direct intervention of a user/designer to record the rationale.

It is also possible to distinguish between "process-oriented" approaches [57], supporting complex system designs and based on Issue Based Information System (IBIS) [58] framework, and "feature-oriented" approaches, used to support automated reasoning [59].

A plethora of approaches, from ontology definition languages [60], to operation-mining algorithms [61] to soft computing techniques [62] has been proposed for enabling the effective capturing of design rationale. The most recent developments in this area includes the Design Rationale editor (DRed) [54], derived from the IBIS concept, which proposes a simple and unobtrusive approach that allows engineering designers to record their rationale as the design proceeds. A different solution is proposed by CoPe_it! [63], which support argumentative collaboration on the Web. The use of automated design rationale capture systems to improve collaborative co-design CAD environments has been also extensively studied [64] and several attempt have been made to capture design knowledge and to represent it in a single-user CAD environment [65].

Knowledge Enhanced Notes (KEN) [66] is a system enhancing the capturing of discursive and collaborative aspects of synchronous design activities. COHERE [46] emerges from work in issue mapping and design rationale, to propose a social, semantic annotation tool focused on the construction and management of connections—between data, knowledge resources (such as documents), ideas (including issues, options and arguments) and people.

3.6.4 Enabling Downstream Knowledge Capturing

When designing a new product or service, it is increasingly important to have a clear picture of a wide variety of customer needs to identify opportunities to improve the current offer both from a product and service perspective. Enhancing the interaction with the end users to exploit their knowledge in the preliminary design activities is a dream for many manufacturers, which the latest achievements in the area of communication technologies can make true.

Crowdsourcing [67] represents nowadays one of the most promising approaches to capture downstream knowledge for the benefit of product development. Crowdsourcing essentially means outsourcing a task to a large group of people or to a community (a crowd) through an open call. The basic principle is that online communities greatly add value to new product development [68], thus Web 2.0 can leverage the critical role that customers and the crowd play in the innovation process [69]. Crowdsourcing is hugely popular in the software development domain. Dell IdeaStorm [70] represents an example of how idea crowdsourcing may be exploited in an early design stage of new products. Several top companies, such as Microsoft, Apple or IBM, have made extensive use of social media to share ideas with lead users ahead of beta testing [71]. In manufacturing, Web 2.0 applications have been used to gather innovations for both products an services, sometime even by means of virtual prototypes online [72] or SecondLife® extensions [73] to analyze patterns of usability and preferences.

3.6.5 Enabling Knowledge Maturity Assessments

In product development, many problems are intrinsically wicked [74] or ill-defined [75], hence making a decision is often about settling for what is good enough rather than waiting for the optimal solution to emerge, e.g. because of flawed or missing information [76, 77]. Rational decision-making [78] is de-facto a rare occurrence. Methods and tools are needed, therefore, to provide decision-makers with a deeper understanding of the status of the knowledge base on which they draw decisions. The knowledge maturity concept is interesting to boost decision-makers' confidence when information and knowledge is in limited supply. Grebici et al. [79] denote maturity as the distance between the actual level of completeness relative to what should be the level of completeness, i.e. as-is status versus to-be status, thus being the relative state of the development of a piece of information with respect to achieving a purpose. For example, a preliminary analysis result concerning the heat tolerance of an aero engine component may be good enough in the feasibility stage of designing the component, whereas the same numbers may be too inaccurate to be of value in the detail design stage.

The concept of knowledge maturity [80] is explored to leverage KEE practises, providing a practical decision support that increases decision-makers' awareness of the knowledge base on which the decisions are based, and to support cross-boundary discussions on the perceived maturity of available knowledge. The knowledge maturity model [80] computes the state of readiness of a knowledge asset using a narrative scale over three dimensions: input, method (tool) and expertise (experience) on a scale from 1–5. A rank as 5 indicates an excellent knowledge maturity, meaning that content and rationale have been tested and proven and reflect a known confidence, that the procedure to produce the content and rationale reflects an approach where proven methods are used, where workers continually reflect and improve, and that lessons learned are recorded. On the other end, a knowledge maturity level ranked with 1 indicates that the content and rationale is characterized by instability (e.g., due to a poor understanding of the knowledge base), that the procedure to produce the content and rationale is dependent on individuals, and that formalized methods are non-existent. In between these levels there is a continuous increase in detailing, documentation and standardization.

The knowledge maturity allows worldwide-located teams to have a shared artifact around which they can identify and discuss issues of concern, visualize the current status of the knowledge base and negotiate a shared understanding of the advantages and drawbacks with the available knowledge elements.

3.7 Conclusion

Engineering work is little-by-little changing to include a larger part of the value creation activities in virtual and cross-functional development processes. At the same time, the definition of 'engineer' is changing as well. Engineering organizations are hiring younger engineers who think and work differently, and who must learn to utilize their knowledge, capabilities and talent in more open and collaborative way.

These aspects raise inevitable questions on the role of humans in product development. Are they engineers? Are they product developers? Are they function providers? Are they knowledge workers? The engineering culture, which changes slowly in established firms, is struggling to adapt to the relatively rapid changes of targets, methods and workforce.

In the new context, engineers are explicitly interested in avoiding redundancy, and instead seek novelty and innovation rather than well-known knowledge. Essentially they are looking for new ways to deal with new problems they are likely to face when moving into the development of complex, lifecycle-oriented solutions in a distributed context. What appears evident from the research is that these new problems are not likely to be adequately addressed by the "traditional" information and knowledge-sharing systems alone. In spite of the hugely important work being done in the CAD/PDM/PLM arena, a quantum leap in methods and tools is required to satisfy the need for knowledge of the new, distributed, digital, cultural diverse, participatory and social responsible generation of engineers.

The research in the area of KEE aims to address such need. The purpose of KEE is twofold: facilitating the work of the "knowledge engineer", the person in a company who is responsible for 'engineering' knowledge bases, as well as the work of the "knowledge worker", of who use such knowledge for accomplishing his/her everyday tasks. In the end, KEE is all about enabling more effective and efficient engineering activities, in an increasingly challenging product development environment, by providing engineers the knowledge they need, in the form they need and at the time they need it.

References

1. Beck P (2005) Collaboration vs. integration: implications of a knowledge-based future for the AEC industry. Design intelligence web. http://www.di.net/articles/archive/2437. Accessed Sep 2009
2. Mathew GE (2002) The state-of-the-art technologies and practises for enterprise business collaboration—an overview. Infosys technologies ltd. http://www.infosys.com/infosys-labs/publications/Documents/TSR_collaboration_1.0.pdf. Accessed Dec 2010
3. Greising D, Johnsson J (2007) Behind Boeing's 787 delays. Chicago Tribune. http://articles.chicagotribune.com/2007-12-08/news/0712070870_1_dreamliner-boeing-spokeswoman-suppliers. Accessed Dec 2010

4. N N (2010) Press centre. Airbus S.A.S. http://www.airbus.com/en/presscentre. Accessed Dec 2010
5. McElroy MW (2003) The new knowledge management: complexity, learning, and sustainable innovation. KMCI Press/Butterworth-Heinemann, Burlington
6. Kern EM, Kersten W (2007) Framework for internet-supported interorganizational product development collaboration. J Enterp Inf Manag 20(5):562–577
7. Davidow WH, Malone MS (1993) The virtual corporation: structuring and revitalizing the corporation for the 21st century. Harper Business, New York
8. Malone TW, Crowston K (1994) The interdisciplinary study of coordination. ACM Comput Surv 26(1):88–119
9. Bordeau J, Wasson B (1997) Orchestrating collaboration in collaborative telelearning. In: du Boulay B, Mizoguchi R (eds) Artificial intelligence in education. IOS Press, Amsterdam
10. Reed CA, Long DP Collaboration, Cooperation and dialogue classification. In: Pollack ME (ed) IJCAI-97: Working notes of the IJCAI'97 workshop on collaboration, cooperation and conflict in dialogue systems, Nagoya, Japan
11. Walton DN, Krabbe ECW (1995) Commitment in dialogue: basic concepts concepts of interpersonal reasoning. State University of New York Press, New York
12. Grice HP (1975) Logic and conversation. In: Cole P, Morgan JL (eds) Studies in syntax, Vol III. Seminar Press, New York
13. Browne J, Zhang J (1999) Extended and virtual enterprises-similarities and differences. Int J Agile Manag Syst 1(1):30–36
14. Ericksen PD, Suri R (2001) Managing the extended enterprise. Purch Today 12(2):1–7
15. Boardman JT, Clegg BT (2001) Structured engagement in the extended enterprise. Int J Oper Prod Manag 21(5/6):795–811
16. Pardessus T (2001) The multi-site extended enterprise concept in the aeronautical industry. Air Space Europe 3(4):46–48
17. Hardwick M, Bolton R (1997) The industrial virtual enterprise. Commun ACM 40(9):59–60
18. Larsson A (2005) Engineering know-who: why social connectedness matters to global design teams. Doctoral thesis. Luleå University of Technology, Luleå
19. Katzy BR, Dissel M (2001) A toolset for building the virtual enterprise. J Intell Manuf 12:121–131
20. Allen T (1977) Managing the flow of technology. MIT Press, Boston
21. Brown JS, Duguid P (2000) The social life of information. Harvard Business School Press, Boston
22. Herbsleb JD, Mockus A, Finholt TA et al (2000) Distance, Dependencies, and delay in a global collaboration. In: SIGGROUP (ed) CSCW 2000: Computer supported cooperative work: ACM 2000 Conference on computer supported cooperative work, Philadelphia
23. Lipnack J, Stamps J (1997) Virtual teams. John Wiley & Sons, New York
24. Brandenburger AM, Nalebuff BJ, Kaplan RS (1995) The right game: Use game theory to shape strategy. Harvard business review (Jul-Aug):57–71
25. Loebecke C, van Fenema PC, Powell P (1999) Co-opetition and knowledge transfer. Data Base Adv Inf Syst 30(2):14–25
26. Appleyard MM (1996) How does knowledge flow? interfirm patterns in the semiconductor industry. Strateg Manag J 17:137–154
27. N N (2009) www.smuts.uct.ac.za/~dliang/essays. Accessed Sep 2009
28. N N (2007) VIVACE Project. http://www.vivaceproject.com. Accessed Jan 2011
29. Sellini F, Cloonan J, Carver E et al (2006) Collaboration across the extended enterprise: barriers and opportunities to develop your knowledge assets? In: Horvath I, Duhovnik J (eds) Tools and methods of competitive engineering-proceedings TMCE 2006 conference. Ljubljana, Slovenia
30. Sandberg M (2005) Knowledge enabled engineering design tools for manufacturability evaluation of jet engine components. Licentiate thesis. Luleå University of Technology, Luleå

31. Isaksson O (2003) A generative modeling approach to engineering design. In: Folkeson A, Gralen K, Norell M et al (eds) Proceedings ICED 03, the 14th international conference on engineering design, Stockholm
32. Nergård H (2006) Knowledge enabled engineering systems in product development: towards cross company collaboration. Licentiate thesis. Luleå University of Technology, Luleå
33. McAfee A (2006) Enterprise 2.0: the dawn of emergent collaboration. MIT Sloan Manag Rev 47(3):21–28
34. Larsson A, Ericson Å, Larsson T et al (2008) Engineering 2.0: exploring lightweight technologies for the virtual enterprise. In: Proceedings of 8th International Conference on the design of cooperative systems. Carry-Le-Rouet
35. Bughin J (2009) How companies are benefiting from Web 2.0: McKinsey global survey results. mckinsey quarterly. http://www.mckinseyquarterly.com/How_companies_are_benefiting_from_Web_20_McKinsey_Global_Survey_Results_2432. Accessed Dec 2010
36. Richards D (2009) Social software/web 2.0 approach to collaborative knowledge engineering. Int J Inf Sci 179:2515–2523
37. Maalej W, Happel HJ (2008) A lightweight approach for knowledge sharing in distributed software teams. In: Yamaguchi T (ed) Practical aspects of knowledge management, 7th international conference, PAKM 2008, Yokohama, Japan, Nov 22–23, 2008
38. Patrick K, Dotsika F (2007) Knowledge sharing: developing from within. Learn Organ J 14(5):395–406
39. Ciavola BT, Gershenson JK (2009) Establishment of an open, Wiki-based online resource for collaboration in the field of product family design. In: Norell Bergendahl M, Grimheden M, Leifer L (eds) Proceedings of ICED'09, Stanford, CA
40. Millen D, Feinberg J, Kerr B (2006) Social bookmarking inn the enterprise. IBM. http://www.research.ibm.com/jam/601/p28-millen.pdf. Accessed Jan 2011
41. Høimyr N, Jones PL (2007) Wikis supporting PLM and technical documentation. In: Schmitt R (ed) Holistic PLM implementation-mastering the organisational change behind the obvious. Eurostep proceedings, Geneva
42. Hawryszkiewycz IT (2007) Lightweight technologies for knowledge based collaborative applications. In: Proceedings of IEEE joint conference on e-commerce technology and enterprise computing, E-commerce and E-services, IEEE computer society, Tokyo
43. N N (2011) Vuuch enterprise social system. http://www.vuuch.com. Accessed Jan 2011
44. Sanghee K, Bracewell RH, Wallace KM (2007) Improving design reuse using context. In: Bocquet JC (ed) Proceedings of ICED 2007, The 16th international conference on engineering design, Paris
45. Kirsch-Pinheiro M, Villanova-Oliver M, Gensel J et al (2005) Context-aware filtering for collaborative web systems: adapting the awareness information to the user's context. In: Liebrock LM (ed) Proceedings of the 2005 ACM symposium on applied computing. ACM, New York
46. De Liddo A, Buckingham Shum S (2010) Cohere: a prototype for contested collective intelligence. Workshop on collective intelligence in organizations. In: Quinn KI, Gutwin C, Tang JC (ed) Proceedings of the 2010 ACM conference on computer supported cooperative work, ACM, Savannah
47. Prekop P, Burnett M (2003) Activities, context and ubiquitous computing. Comput Commun 26(11):1168–1176
48. Gustavsen RM (2002) Condor—an application framework for mobility-based context-aware applications. In: Ljungstrand P, Holmquist LE (eds) Adjunct proceedings UBICOMP 2002. Göteborg
49. Redon R, Larsson A, Leblond R et al (2007) VIVACE context based search platform. In: Kokinov B, Richardson DC, Roth-Berghofer ThR et al (eds) Modeling and using context: 6th international and interdisciplinary conference, CONTEXT 2007, Roskilde
50. Heisig P, Caldwell NHM, Grebici K et al (2010) Exploring knowledge and information needs in engineering from the past and for the future e results from a survey. Des Stud, doi: 10.1016/j.destud.2010.05.001

51. Hooey BL, Foyle DC (2007) Requirements for a design rationale capture tool to support NASA's complex systems. In: Proceedings of 3rd international workshop on managing knowledge for space missions. Pasadena
52. Klein M (1993) Capturing design rationale in concurrent engineering teams. IEEE Comput J 26(1):39–47
53. Bracewell R, Gourtovaia M, Moss M et al (2009) DRed 2.0: a method and tool for capture and communication of design knowledge deliberated in the creation of technical products. In: Norell Bergendahl M, Grimheden M, Leifer L (eds) Proceedings of ICED'09, Stanford, CA
54. Gross MD (1996) The electronic cocktail napkin–computer support for working with diagrams. Des Stud 17(1):53–69
55. Shipman FM, McCall RJ (1997) Integrating different perspectives on design rationale: supporting the emergence of design rationale from design communication. Artif Intell Eng Des Anal Manuf 11(2):141–154
56. Hu X, Pang J, Pang Y et al (2000) A survey on design rationale: representation, capture and retrieval. Proceedings of ASME DETC/CIE 2000, Baltimore, MD
57. Conklin EJ, Yakemovic KCB (1991) A process-oriented approach to design rationale. Hum Comput Interact 6(3–4):357–391
58. Goodwin R, Chung PWH (1997) An integrated framework for representing design history. J Appl Applied Intell 7(2):167–181
59. Chandrasekaran B, Goel AK, Iwasaki Y (1993) Functional representation as design rationale. IEEE Comput Soc Press 26(1):48–56
60. Medeiros A, Schwabe D, Feijó B (2005) Kuaba ontology: design rationale representation and reuse in model-based designs. Lect Notes Comput Sc 241–255
61. Jin Y, Ishino Y (2006) DAKA: design activity knowledge acquisition through data-mining. Int J Prod Res 44(14):2813–2837
62. Saridakis KM, Dentsoras AJ (2007) Soft computing in engineering design-a review. Adv Eng Inform 22(2):202–221
63. N N (2010) CoPe_it!. http://copeit.cti.gr/site/index.html. Accessed Jan 2011
64. Ryskamp JD, Jensen CG, Mix K et al (2010) Leveraging design rationale to improve collaborative co-design CAD environments. In: Horváth I, Mandorlini F, Rusák Z (eds) Proceedings of TMCE 2010 symposium. Ancona
65. Sung R, Ritchiea JM, Reab HJ et al (2010) Automated design knowledge capture and representation in single-user CAD environments. J Eng Des doi:10.1080/0954482090 3527187
66. Conway A, William I, Andrew L (2009) Knowledge enahnched notes (KEN). In: Norell Bergendahl M, Grimheden M, Leifer L (eds) Proceedings of ICED'09, Stanford, CA
67. Howe J (2006) The rise of crowdsourcing. wired magazine. http://www.wired.com/wired/archive/14.06/crowds_pr.html. Accessed Jan 2011
68. Pitta DA, Fowler D (2005) Online consumer communities and their value to new product developers. J Prod Brand Manag 14(5):283–291
69. Ribiere MV, Tuggle FD (2009) Fostering innovation with KM 2.0 VINE. J Inform Knowl Manag Sys 40(1):90–101
70. Di Gangi PM, Wasko M (2009) Steal my idea! organizational adoption of user innovations from a user innovation community: a case study of dell ideastorm. Decis Support Sys 48(1):303–312
71. Smith D, Valdes R (2005) Web 2.0: get ready for the next old thing. Gartner research paper, Stamford, CT
72. Füller J, Bartl M, Ernst H et al (2006) Community based innovation: how to integrate members of virtual communities into new product development. Electron Commer Res 6:57–73
73. Tseng MM, Jiao RJ, Wang C (2010) Design for mass personalization. CIRP Ann Manuf Technol 59:175–178
74. Rittel HWJ, Webber MM (1973) Dilemmas in a general theory of planning. Policy Sci 4:155–169

75. Cross N (1982) Designerly ways of knowing. Des Stud 3(4):221–227
76. Flanagan T, Eckert C, Clarkson PJ (2007) Externalizing tacit overview knowledge: a model-based approach to supporting design teams. Artif Intell Eng Des Anal Manuf 21:227–242
77. Rosenzweig P (2007) Misunderstanding the nature of company performance: the halo effect and other business delusions. Calif Manag Rev 49(4):6–20
78. Simon HA (1997) Rational decision making in business organizations. Am Econ Rev 69(4):493–513
79. Grebici K, Goh YM, Zhao S et al (2007) Information maturity approach for the handling of uncertainty within a collaborative design team. In: Bannon L, Wagner I, Gutwin C et al (eds) Proceedings of the 10th european conference on computer supported cooperative work. Limerick
80. Johansson C (2009) Knowledge maturity as decision support in stage-gate product development: a case from the aerospace industry. Doctoral thesis. Luleå University of Technology, Luleå

Chapter 4
Evolution in Mechanical Design Automation and Engineering Knowledge Management

Giorgio Colombo and Ferruccio Mandorli

Abstract Design activity consists, strictly speaking, in synthesizing something new (or arranging existing things in a new way) to satisfy a recognized need. This activity is accomplished through an iterative, knowledge-based, decision-making process. One of the goals pursued by IT applied to product development has been Design Automation that is the execution of all tasks of the design process of a product by a software application. In this context, engineering knowledge elicitation, representation and management are key issues to achieve the so-called "right the first time" design. This chapter summarizes authors research in Mechanical Design Automation (MDA) domain during the last two decades; it focuses the evolution from CAD centered applications to Object-Oriented (O-O) ones up until the more recent issues related to storing, sharing and reusing of engineering knowledge, the role in PLM approach and the support to decision-making. Finally, the fundamental aspects related to development of real-MDA applications are discussed, on the basis of personal experiences of the authors.

4.1 Introduction

During their design activities, mechanical designers must have the skill to synthesize into a final solution, conflicting issues related to required functions, product shape, manufacturing techniques and materials and economic constraints.

G. Colombo (✉)
Politecnico di Milano, Milano, Italy
e-mail: giorgio.colombo@polimi.it

F. Mandorli
Università Politecnica delle Marche, Ancona, Italy
e-mail: f.mandorli@univpm.it

M. Bordegoni and C. Rizzi (eds.), *Innovation in Product Design*,   55
DOI: 10.1007/978-0-85729-775-4_4, © Springer-Verlag London Limited 2011

The design process, as well as the production processes, has been deeply studied during the past: the traditional action-based representation of design, based on a sequence of linked activities, has been changed in a product-centered representation, able to convey new paradigms like concurrent engineering and co-design.

In the last decade, other product-related aspects, like distribution, maintenance, disposal and recycling, came up to the design and production and, together with the use phase, lead to the concept of product life cycle.

As this evolution took place, the product development process shifted from a production-driven process to a design-driven process, based on engineering knowledge management.

In this context, Design Automation (DA) has played an important role to effectively innovate and improve design procedures; in industrial engineering we can more properly speak of Mechanical Design Automation (MDA).

MDA can be defined as a set of methods, tools and applications that permit to automate the design process; it can be applied to all the phases of the process, from the conceptual to final one related to the production of technical documentation. However, in our opinion, MDA well fits the phases of parametric and detailed design during which a technical solution is completely formalized. In particular, more important benefits from a MDA application can be achieved when dealing with products and parts characterized by a well-defined architecture and design process. Some examples of products falling within this category are heat exchangers, gears, machine tools, industrial mixers and so on. For these types of products, a completely automated procedure can be implemented and the configuration of a new product can be executed by a software application.

DA has a history that matches with that one of IT; first applications were developed with general purpose programing languages and concerned specific aspects of design process such as kinematic analyses or structural calculation. A remarkable improvement was gained when CAD techniques were developed; in fact, graphics and geometric modeling play a fundamental role in product design. The integration between CAD models and programing languages permitted the development of automatic procedure to configure parts and simple products; there are recent examples of this approach [1]. This approach was enhanced by the development of parametric CAD [2]; a lot of applications to configure parts and products have been developed by using parametric models and programable tools such as spreadsheet. In SMEs (Small Medium Enterprises) this is actually one of more diffuse method to develop simple DA applications.

At the same time, a number of methods and tools derived from Artificial Intelligence appeared; in particular, Knowledge-Based Engineering (KBE) tools based on Object-Oriented (O-O) approach, which constitute also today a valid approach to DA. They had a relevant impact within aeronautical and automotive companies; some industrial applications have been developed in these contexts [3, 4]. Side-by-side, researches on methods to acquire and formalize knowledge have been started; in fact, knowledge formalization is a fundamental issue, and several efforts [5] have been carried out in this direction. KBE tools evolved from initial elementary implementation to the actual one characterized by simple programing language, powerful tools to define customized GUI and to integrate external programs (CAD systems, FE solvers, spreadsheet, data base, etc.).

In this chapter we present our experiences in the last two decades in developing real industrial DA applications and we discuss topics in this context and in Engineering Knowledge Management (EKM) in general. In detail, we present and discuss the evolution from first applications centered on parametric CAD integrated with programing languages or calculus tools, then on ones based on Object-Oriented approach and finally the new approach based on agents assembled to carry out all the design tasks.

Moreover, attention is also focused on engineering knowledge; it is very complex, differentiated and interrelated, involving explicit and implicit information, often based on tacit assumptions. A key aspect in the development of MDA applications is the elicitation and the formalization of the design knowledge and the representation of such knowledge into the design support system. This issue is at higher level than ones of MDA and will determine functionalities of IT systems finalized to design process.

4.2 A Brief History of Our Works in MDA and EKM

As mentioned, the history of DA and EKM matches practically with that one of IT; first applications, finalized to mechanism or structural analyses, focused specific problems without considering the complete process; only in recent years attention is focused on a global approach to design.

CAD techniques development represented a milestone in MDA; we highlighted the fundamental role played by graphics and geometric modeling in design process. In particular, parametric CAD has represented, and represents also today, a powerful tool to implement MDA. The first our contribution in DA domain date from 1989; in [2] Colombo et al., proposed an approach to represent design rules directly related to parametric drawing of mechanical parts. Innovations presented in that paper were: a 2-D parametric CAD system based on representation of geometric entities and constraints in a graph, a spreadsheet-like tool to manage geometric and functional parameters and a simple programing language to define sizing or analyses rules. A lot of applications to configure parts and products have been developed by using parametric models and programing languages or tools such as electronic spreadsheets. Today, this approach is the most used in SMEs (Small Medium Enterprises) to develop MDA applications; the evolution, in this case, is not in the approach but in the better tools functionalities.

At the same time, methods and tools developed in Artificial Intelligence domain were applied to Engineering Design; in particular, O-O programing furnished a great contribute. The first KBE (Knowledge-Based Engineering) development systems were based on an O-O programing language [6]. We approached this technology and we realized some prototypes by using a system named The

Concept Modeler® from Wisdom Systems. In [7] we presented a MDA application to configure a family of shearing dies starting from the drawing of the punch sheet. In this application significant contributions were the O-O representation of the product structure and the use of a graphic language proposed by Coad and Yourdon [8] to synthesize object, parameters and methods implemented. This last aspect is important in EKM; UML was further developed starting from this first approach to the problem of representing object, data and relations.

A first discussion in differences between approaches to MDA based on CAD models and programing languages or tools and O-O programing was presented in [9].

In the last decade, we concentrated our attention to dissemination of MDA technologies and we developed some applications by using different tools, which can be referred to the two previously cited approaches. Some methodological aspects are presented in [10, 11]; while examples of implementation with different KBE tools are summarized in [12]. Susca et al. in [13] presented an application to automatically calculate mass properties of a racing car, developed by using the KBE tool named Selling Point® by Oracle. Just to the new century, MDA was applied mainly in aerospace and automotive companies, i.e. industrial contexts with relevant human resources and know-how (a selection of KBE publications produced by the Design of Aircraft and Rotorcraft group at Delft University is collected in the booklet "Knowledge-Based Engineering Supported Design", available on the DAR web site).

In the last years our attention was addressed to implement MDA in SMEs, and, moreover, we identified issues of MDA: product modeling, process modeling, integration with PDM/PLM, knowledge reusability and sharing and support to decision-making. Examples of MDA applications developed for SMEs are reported in [14, 15].

Experiences in developing MDA in SMEs were synthesised in a methodology named MEDEA, which considers all the tasks of the development of an application and proposes methods and tools for each of them [16]. The issue of the integration with PLM/PDM systems was discussed in [17]; in particular, the part coding was used to retrieve existing components in company data base. In various contexts we highlighted relevance of product and process representations; examples are in [16] and [18]. An example of knowledge reusability and sharing was presented in [19].

In the following sections, by using examples, we will clarify our approach to MDA and the results obtained in our work.

4.3 Methods and Tools for Mechanical Design Automation

The overall objective of MDA is to develop computerized systems able to support the engineer in performing his/her design-related tasks, generally within the context of product or process design.

Although MDA systems can be very different to each other, they are usually generative systems, i.e. they use a set of pre-defined procedures to automatically (or semi-automatically) generate different outputs starting from different inputs. Fully-automatic systems just require to the user to specify the inputs; semi-automatic systems can require additional interactions with the user in order to solve special cases and exceptions, not covered by the pre-defined procedures.

Generally speaking, the inputs are the design specifications; the generative procedures are the design rules and the outputs are the design solutions.

In order to implement such kind of systems, both methodological and technological issues must be addressed and solved.

From the methodological point of view, the issue concerns the definition of a set of appropriated procedures and guidelines to identify, to capture, to formalize and to maintain the knowledge related to the problem under inspection.

From the technological point of view, the issue concerns the definition of a set of tools and functions suitable to implement and manage the data structures and the procedures used to represent the design process knowledge.

Several methodologies to support MDA implementation in real-industrial context have been proposed: some of them are particularly suitable for the DA implementation in a Small/Medium Enterprise (SME), others have been proposed by consortium of large companies.

A common feature of all the proposed methodologies is the O-O approach to the problem analysis: the product (or process) to be designed is analyzed and represented as a hierarchical structure made of objects that are defined by means of a set of properties. The instantiation of the properties values define a specific design solution. The procedures to compute the properties values represent the design rules.

The MEDEA (Methodology per Design Automation) methodology focused the attention on some characteristics of the design process in SMEs [16]. It proposes a step-by-step roadmap and suggests methods and tools finalized to developers more skilled on products and design process than on IT technologies. It is based on five main steps:

- *Specs definition*: identification of DA application specs and the criteria to make re-usable and sharable blocks of the product and process data;
- *Knowledge acquisition*: collection of the knowledge related to the product architecture and the design process;
- *Knowledge formalization*: representation of product architecture (tree diagram and UML class diagram [20]) and of the process model (IDEF0 and IDEF3 diagrams [21]);
- *Integration with PDM/PLM system*: definition of the interactions between DA application and company's PDM/PLM system;
- *Implementation of DA application*: using a KBE system.

An alternative methodology has been proposed as the result of the EU project MOKA "Methodology and tools Oriented to Knowledge based Applications" [5].

The proposed methodology focuses on knowledge formalization and data structuring, mainly from an implementation point of view.

MOKA identifies four key technical roles in developing DA applications: the domain experts (they provide main source of knowledge and specification); the knowledge engineers (players that orchestrate the capturing and formalizing activities); the software engineers (the people who builds software, based on requirements set by knowledge engineers); the end users (the ultimate customers of the application). The MOKA roadmap defines the development paths and who does what, when, and what to do when things go wrong.

Additional proposed methodologies are: KOMPRESSA (Knowledge-Oriented Methodology for the Planning and Rapid Engineering of Small-Scale Applications) and DEKLARE (Design Knowledge Acquisition and Redesign Environment) [22].

From the implementation point of view, the issues are related to the definition of appropriated data structures and algorithms to represent and manage the product/process model as well as the implementation of the procedures to compute the model properties.

Although MDA can be very different from each other, some common features and functionalities can be identified: the product/process model is represented as a hierarchical structure made of parts and sub-parts, defined by means of a set of properties, related to each other in a parametric way.

The properties are used to store all the different types of information that are relevant to consolidate the design knowledge into the product model. This information can be stored in form of explicit knowledge (i.e. properties values, like standard dimensions, materials, physical properties, etc.) or procedural knowledge (i.e. procedures to compute the properties values, like dimensioning algorithms, configuration and compatibility rules, check and analysis procedures, etc.).

In order to manage such kinds of models, the system must incorporate some dependency backtracking or history-based mechanism to keep updated the properties values.

DA systems are focused on knowledge capture and on the definition and implementation of design procedures. In the mechanical field, it often happens that a significant part of the information to be managed during the analysis and synthesis design phases is related to components shape, position and dimensions. For this reason, it is quite common that a significant module of a DA system is the geometric kernel.

Different approaches and IT tools can be used to implement DA applications and the last two decades have shown the impact of IT evolution on DA applications development. In the following, we summarize the three main approaches for the development of DA applications, based on different technologies: CAD centered approach, KBE approach and mixed approach.

The CAD centered approach is particularly suitable to develop DA applications where the largest part of the information to be managed is somehow related to the geometric properties of the product model. The CAD model, enriched with appropriated customized properties, represents the product model. The use of parametric CAD model templates allows the definition of generic models. Traditional programing languages can then be used to implement algorithms that, by using the functions provided by the CAD Software Development Kit (SDK), allow to instantiate the parameters stored into the model templates, in order to generate a specific product model configuration.

The KBE approach is particularly suitable to develop DA applications where geometry is one of many kinds of information to be managed. By following this approach, 2-D or 3-D models are usually outputs, automatically generated by the system. It is based on the use of a KBE development framework. These frameworks are based on a simplified O-O programing language and provide functionality to handle the hierarchical product model structure, including procedure to generate shapes, to set dimensions, to apply assembly rules, to access to external data bases, etc.

Finally, the mixed approach is particularly suitable when the generative methodology, provided by the KBE framework, needs to be coupled with the interaction provided by the use of traditional CAD systems. The main feature of a KBE development framework able to support this approach is that the geometric kernel it is not embedded into the framework, but the framework provides functions to interface and drive an external CAD system.

Theoretically speaking, the implementation of a DA system can be considered a technicality. However, in a real-industrial context, the selection of the most appropriated approach and developing tool is directly related to aspects such as (1) the capability to foresee the effort and the expertise required to develop, to test, to validate and to maintain the system; (2) the capability to predict the trade-off between development costs and use benefits; (3) the possibility for the domain expert and/or the knowledge engineer to directly participate to the system implementation and (4) the possibility for the domain expert to maintain and upgrade the system, by-passing the software engineer and/or the knowledge engineer.

In the following we introduce some examples of applications developed adopting the mentioned approaches.

4.4 DA Applications Based on CAD Models and Software Components

The application example reported in this section refers to a KBE application developed to support the design and production of gas turbine ducts. The research was carried on in collaboration with an Italian company, supplier of General Electric.

As shown in Fig. 4.1, the input of the ducts design process is the 2-D drawing of the draft layout of the plant. The results of the design activity will be: the 3-D detailed design of the ducts, the 2-D drawings required for the ducts production, the bill of material (BOM) and the cost estimation of the ducts.

Fig. 4.1 Tasks share

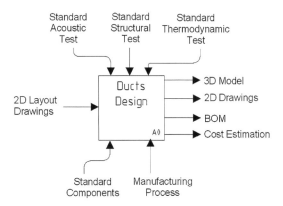

During the design activity, the designer is constrained by economical considerations (i.e. the use of standard components and semi-finished materials), technological aspects (i.e. the manufacturing processes) and limits imposed by
standard regulations (i.e. acoustic, structural and thermodynamic tests).

The traditional ducts design is a trial and test process: the preliminary design
solution is obtained by mapping the generic ducts parts, reported in the 2-D draft
drawings, into standard ducts modules, on the basis of the designer experience.

The preliminary solution is, then, tested by means of the thermodynamic,
acoustic and structural tests imposed by standard regulations and design specifications. If the tests are not passed, the design must be modified and then the tests
need to be re-executed.

When the final design solution is achieved, all the documents required for the
production must be compiled, including BOMs made of several thousands of
components.

The efficiency of the described process is clearly affected by three main constraints: the capability of the designer to rapidly converge to a sound solution; the
time required to validate the solution (i.e. to perform the required tests); the time
required to compile the production documentation (i.e. drawings and BOMs).

In the described context, the objectives of a KBE design support system will be:
to capture and to consolidate the multi-disciplinary know-how of the designer, in
order to let the system generate sound design solutions (from the manufacturability
and assimilability point of view); to automate as much as possible the repetitive
and time-consuming tasks, in order to shorter the time required to validate the
generated design solution and to compile the production documents.

The design knowledge elicitation started with the functional analysis of several
plant typologies. This analysis allowed identifying the modularity of the systems
and enabled the synthesis of the different types of plants into a default template
and a set of rules to manage the product variants.

The plant template was then represented with a hierarchical model structure by
using an O-O approach. Such model contains all the information and rules required
to generate the detailed geometric configuration of the different parts of the ducts.

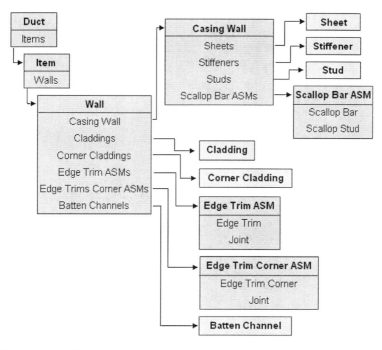

Fig. 4.2 Part of the duct hierarchical structure

Figure 4.2 shows part of the hierarchical structure of the duct: the duct is an ordered set of modules, called items, along a 3D mean axis made of line segments. Each item is generally made of four walls. A wall is made of the structural part, the casing wall and insulation. Casing wall consists on sheet metal panels, some stiffeners and parts to support insulation (studs and scallop bars). The insulation is made of claddings, edge trims and batten channels which all are sheet metal parts that contain insulation fibers.

From the methodological point of view, the results of the knowledge elicitation phase have been logically organized by using formal structures like Configuration Virtual Prototypes (CVP) and Design Structure Matrix (DSM), as explained in [23]. In particular, the DSM approach has been used to identify and relate to each other the functional requirements, as the CVP model has been used to define the overall product structure.

From the implementation point of view, the system has been developed integrating in house developed algorithms and data structures with data structures provided by a commercial 3-D MCAD system, a relational database and GUI manager. The user interacts with the system at different levels and at different stages of the design procedure.

The overall structure of the duct is defined by the user thanks to a table like input interface, where he/she can specify number, type, dimensions and optional components of each module of the duct, as shown in Fig. 4.3.

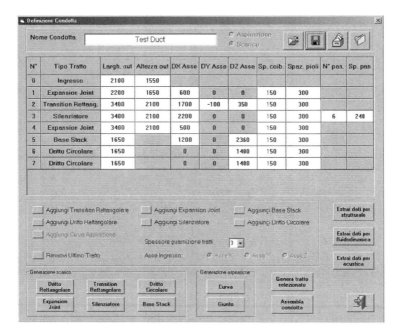

Fig. 4.3 Duct modules data entry

Fig. 4.4 Automatically generated 3-D duct model

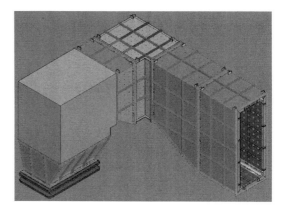

The system takes few minutes to automatically generate the detailed 3-D model of the duct from the data specified by the user and the pre-defined parametric CAD model templates. As an example, the model shown in Fig. 4.4 is made of about 5.000 parts.

If particular exceptions need to be managed, the user can directly access the CAD model in order to make local modifications.

Suitable procedures are then charged to access both geometrical and non-geometrical information in order to extract the data required to perform

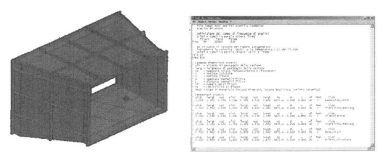

Fig. 4.5 Geometrical and non-geometrical data automatically extracted from the duct model

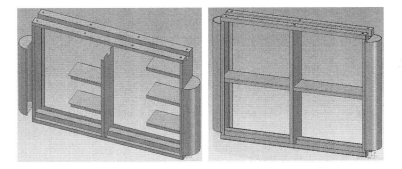

Fig. 4.6 Example of assembly template (*left*) and assembly instance (*right*)

acoustic, thermodynamic and structural analysis (see Fig. 4.5) as well as to generate production documents.

It is well known that a critical issue related to the development of KBE application is the system maintenance and upgrade. In order to face this issue, the adopted approach is based on the wide use of parts templates (parametric CAD models) linked to part data (e.g. code, standard dimensions, material, etc.) stored into the database.

The definition of a new component requires the interactive modeling of the geometric template and the input of the data into the database.

In a similar way, assembly templates and database data are used to represent the hierarchical structure of the ducts. The positioning of the parts into the assembly is managed thanks to the traditional mating relationships available in any MCAD system and thanks to the use of 2-D sketches.

The use of assembly templates and sketches is summarized by the two following examples.

The left part of Fig. 4.6 shows the template of the silencer panel framework. When a specific instance of the template needs to be generated, the assembly algorithm traverses the assembly template and collects component references and mating relationships. The final assembly is then automatically generated by

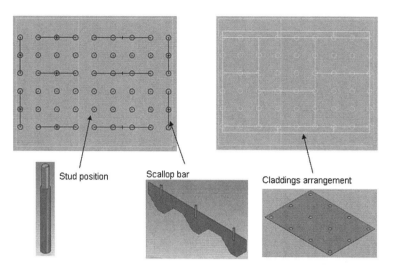

Fig. 4.7 Example of 2D sketches to drive the automatic assembly procedure

checking the components that must be present, by generating the models of the parts to be included in the assembly and automatically applying the mating constraints (as shown in the right part of Fig. 4.6).

Simplified layouts using 2-D sketches can be successfully used to add flexibility in the management of part positioning.

In case of insulation panel definition, the basic components to be assembled are studs, scallop bars and claddings (see lower part of Fig. 4.7).

Studs are characterized by revolution geometry and are welded on panels. Then, only points are necessary for their position: circle centers will represent those points. Scallop bars are represented by segments which, combined with stud circles, provide all necessary geometrical parameters. Claddings are represented by closed paths passing through stud centers.

The insulation panel layout can then be defined by using 2-D sketches where different types of 2-D geometrical entities are used to identify the component type and position (see upper part of Fig. 4.7).

The default layout can be computed on the basis of known rules and specific panel dimensions. The sketch corresponding to the default layout is then automatically generated by the system when an insulation panel needs to be instantiated. If a different layout is required, the user can interactively modify the 2-D sketch and ask the system to re-compute the configuration.

The automatic procedure analyzes the modified sketch and extracts the information for the instantiation and generation of the parts and the subassemblies. Finally, the components are assembled to their final configuration, as shown in Fig. 4.8.

The adopted approach has proven to be a good compromise between a fully automated procedure and the need to manage highly-configurable product variants.

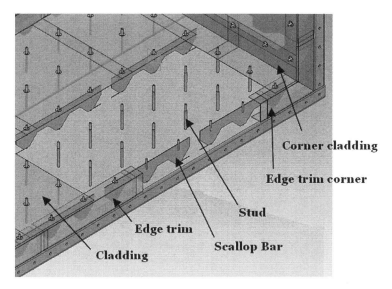

Fig. 4.8 Example of instantiated assembly

4.5 DA Applications with KAE Tools

This section describes two industrial applications developed adopting the MEDEA methodology. The first is related to the configuration of fired heaters, while the second concerns an industrial mixer and is focused on process representation and integration with PDM system.

4.5.1 Fired Heater Configuration

The term "product configuration" means different activities that can be oriented to marketing or to product development that respectively refers to the commercial configuration and to engineering or technical configuration. The first application concerns the second case and, in particular, to an Engineer-To-Order (ETO) approach, where the customer needs highly-customized products, such as ships, dedicated machines or industrial plants. In this case, company staff has to understand and translate the customer's needs using an engineering configuration system able to generate configurations of the product and of its components that satisfy initial requirements. In such a context, it has been implemented a knowledge-based (KB) application to automatically configure fired heater for a chemical plant [24].

Fig. 4.9 Petroleum refinery
plant

Fig. 4.10 Fired heater:
a Box, **b** Cabin and
c Cylindrical

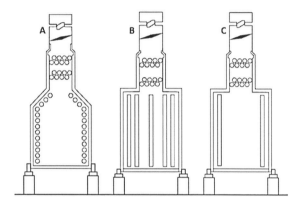

The fired heater is a sub-system of petroleum *refinery plants* used to heat and
partially vaporize oil in order to separate it from the hydrocarbons. Figure 4.9
shows an image of a chemical plant and of a fired heater.

A fired heater consists of a *radiant* section whose heat is transmitted to fluid
flowing inside the tube coils by radiation, a *convection* section where flue gases
leaving the radiant section give up their heat by convection and then released in the
environment through the *stack*.

Generally, more common fire heaters can be grouped in (Fig. 4.10): box, cabin
and cylindrical.

This classification is based on the amount of heat absorbed by the fluid to be
warmed up. This value permits to determine the most convenient solution. Normally, vertical cylindrical type is adequate up to 20–25 106 kcal/h; while cabin or
box for higher values.

According to MEDEA methodology, knowledge acquisition for both product
and process and following formalization have been two important phases for the
development of the KB application.

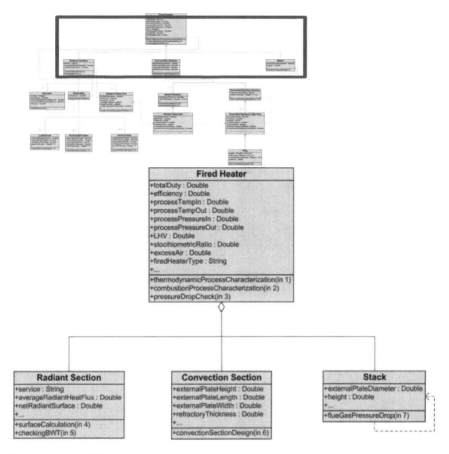

Fig. 4.11 Main UML class diagram

Knowledge has been acquired using different sources: interviews with technical staff, company technical manuals, standards [25, 26] and scientific books.

For the formalization step, different techniques and models have been adopted. First the product structure has been represented as a tree where it is possible to see the mentioned subsystems and their main sub-components; then, it has been translated into UML Static Class Diagrams where subsystems, components and relationships are properly represented. Figure 4.11 portrays the main UML diagram where the three mentioned subsystems (radiant, convection and stack) are represented.

Process knowledge has been formalized adopting IDEF0 and IDEF 3 Process Flow (PF) models (www.idef.com). The design process of a fired heater is based on two inter-related principal activities, partially carried out in parallel:

- Parts arc dimensioned applying design rules derived from the thermodynamics;
- Designer's choices and calculated parameters values are verified on the base of mechanical and fluid dynamics principles.

Design rules that permit both to size correctly each component and generate the product configuration have been associated to the IDEF0 and UML diagrams. Rules have been derived both from thermodynamics principles and designers' experience. As an example, consider activity "Characterize thermodynamic process" or UML class Fire heater, related design rules to calculate heat fired are:

- HFired = Duty Tot/η
  where:
  HFired: = Heat Fired [MW];
  DutyTot = Total Duty [MW].

- HLosses = HFired . ηLosses
  where:
  HLosses = Heat Losses [MW];
  ηLosses = Efficiency Losses.
  It is 2.5% when a pre-heated system is used otherwise it is 1.5%

- ηtot = η + ηLosses
  where:
  ηtot = Total efficiency;
  η = efficiency.

The DA application has been realized using the commercial KBE system, RuleStream.

Figure 4.12 portrays the product tree structure (representing its architecture) where parts are ordered according to assembly and sub-assembly logic and to the UML diagrams. Therefore, a set of elements linked by rules composed the structure.

To implement the design process, it has been necessary to define the rules and relationships among the parts and link them to the associated parameters. All rules related to the design process (formalized with IDEF diagrams) have been implemented linking to each part properties a value derived from an analytical formula or geometric relationship.

The design of a fired heater, such as many other products, requires the execution of specific design activities and in parallel the definition of the product architecture. The KB application has been based on two separate models: the first for the plant components while the second for the configuration/design process. Therefore, a communication channel to transfer data and information from the process to the product and vice versa has been implemented.

The realized application automatically executes the configuration process, performing all activity usually carried out by the designer. It starts with the acquisition of design specs and subsequent steps are executed in accordance with the implemented process.

In collaboration with the technical staff of the involved company the prototype has been tested with two test cases:

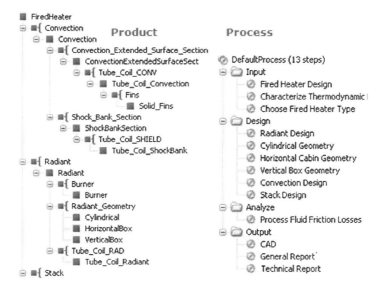

Fig. 4.12 Implemented product structure and design process

- A cabin fired heater with horizontal tubes that can generate a total duty equal to 11.6 MW. It was already developed by the company and the goal has been to demonstrate that the final results generated by the automatic configurator are the same of those obtained following the traditional process within the company.
- A cylindrical and a box fired heater (both able to generate a total duty equal to 10 MW) applying the same data input in order to demonstrate the possibility to rapidly generate and compare different design solutions.

Figure 4.13 portrays the acquisition of customer' requirementss while Fig. 4.14 portrays the configuration obtained for cylindrical fire heater (Fig. 4.14a) and for a box one both with vertical tubes (Fig. 4.14b).

All data generated, from the 3-D CAD models to the final documentation summarizing fire heater characteristics, were in accordance with those calculated by the company.

This permitted to demonstrate either the accordance of the obtained results with those obtained with the traditional process or the possibility to rapidly generate and compare different design solutions.

Company staff considered DA application an optimal tool especially during order acquisition to present the offer to the potential customer and generate technical drawing of the plant, named general arrangements. In this case the introduction within the company could be done in a short time. On the other hand, the introduction of such a technology in the technical departments requires deeper changes and could be done at medium long term.

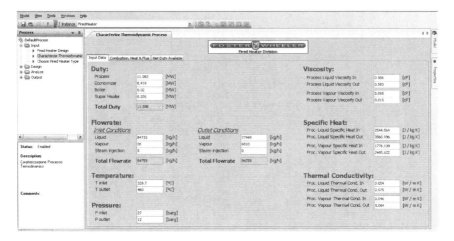

Fig. 4.13 The KB application: entering design specs for thermodynamic characterization

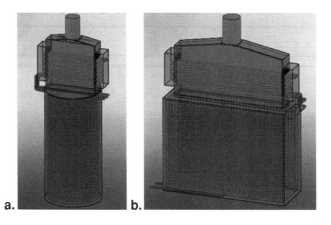

Fig. 4.14 Automatic fired heater configuration: **a** Cylindrical and **b** Box

4.5.2 Integration KBE-PDM

This case study regards the design of a family of industrial mixers and is focused on the integration between KBE and companies' data repositories (PDM system) to improve knowledge reuse in mechanical companies [16, 17].

The design process is addressed to produce 3-D models, drawings and BOM, with updating the company PDM system. This leads to a partial process re-engineering to fulfill the complete integration between KBE and PDM. The designer searches in the PLM/PDM models/documents already realized and tries to reuse it, to avoid the redesign of existing components or products.

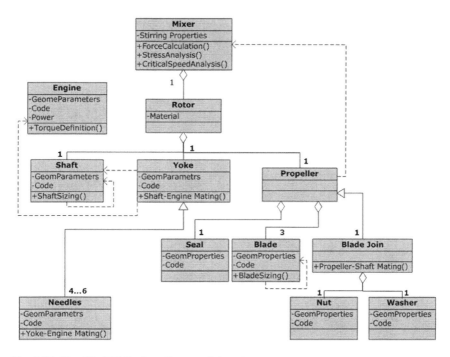

Fig. 4.15 Simplified UML class diagram of the rotor group

As in the previous case, before implementing the application with the KBE kernel, knowledge acquisition and formalization are needed. Product tree structure and UML static class diagrams have been developed to describe relationships among parts and properties. The system is composed of three main sub-systems: engine, rotor and stator. Figure 4.15 describes the rotor group, the relation with the engine and the properties of the top block.

As mentioned, for this test case it has been also necessary to study the inter-action among DA application, PDM system and end-users. Accordingly to MEDEA, it has been described with an UML Activity Diagram (Fig. 4.16), which represents the sequence of the steps necessary to complete the configuration and archiving of a new mixer.

Also in this case we adopted Rulestream since it also enables designers to share their knowledge and reuse their applications or part of them in other applications. The source code of Rule-Stream applications lets the designers personalize their applications, thus permitting the development of a module to retrieve data and documents stored into the company's PDM.

Figure 4.17 shows a snapshot of the application while Fig. 4.18 the functional scheme.

The application allows the designer to automatically design and configure an industrial mixer. It can be used in two different ways:

Fig. 4.16 Operations to
configure the product with
new parts storing

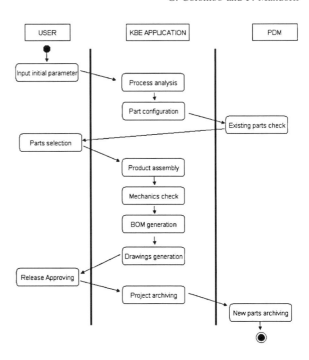

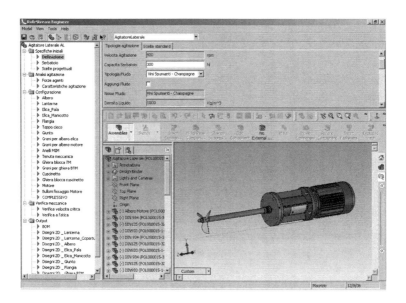

Fig. 4.17 User Interface for the mixer DA application

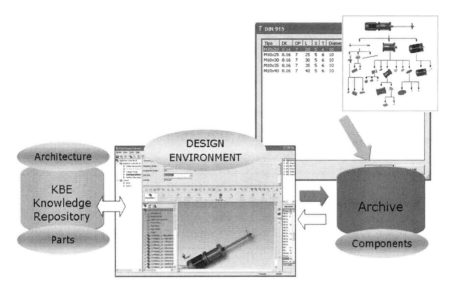

Fig. 4.18 Application functioning scheme

- To configure a new product following all steps defined to design a mixer, such as the calculation of the forces acting on the blades, the sizing of the shaft and so on.
- To query the PDM to acquire product components already developed, using their drawings, codes and data in general. The application generates the code for every component by which it can query the PDM and discover if a part with the same code already exists. On positive case, it returns the component and its drawings and avoids the next design and review steps; otherwise, it lets the designer decide whether to proceed to design a new component with that code, or to investigate into the PDM to find a similar component.

The coding system adopted was similar to that one used in the involved company but simplified according to the nature of DA prototype. In fact, the company uses a coding system that describes very particularly the components, reaching a very high level of description. Anyway, the authors think that using a more simplified coding system does not affect the validation of the prototype.

The validation phase with various case studies and the direct involvement of SMEs' technical staff permitted to verify the effectiveness of the implemented DA application. Meaningful result has been the reduction of the product development time. Typically, within the SME that provided the study case of the heat exchanger, the product design requires about 30 h of an expert designer. This development time is mainly due to the frequent iterative cycles required by some steps of the design process even if they use electronic datasheets for some tasks. Instead, the developed application takes about 15 min ensuring also the respects of standards and design rules, thus quality of the product.

4.6 Discussion and Conclusions

This chapter has presented authors' contribution to the domain of MDA; their experiences started from first examples based on parametric CAD integrated with programing languages or calculus tools to those based on an O-O approach. In last two decades, more interesting contributions to real-industrial MDA are referable to these two methodologies. Other tools have been tested, such as rule-based systems or soft computing, but their relevance in industrial context was very light. Our experience demonstrates that if the basic methods are not changed, the complexity of the applications is really increased. We highlighted the key issues of the current MDA applications: product and process modeling, PLM integration, knowledge sharing and reuse.

CAD and O-O approaches focus on product modeling. We think that this is a good approach for products without high complexity that do not require numerous and articulated tasks to reach the final solution. In fact, only in these cases it is easy to embed design process in parameters and methods of the objects representing the parts. For complex situations, for example integrating numerical simulation, the design process is the kernel of the MDA application. We are not saying that the product model is not necessary, but it is necessary to represent design process to accurately control design choices.

The problems of integration within PLM/PDM solutions, sharing and reusing knowledge are strictly related and very complex and we think that MDA methodologies must play the fundamental game for its own future. Practically, all the MDA applications we have developed are "black box", i.e. they encapsulate all knowledge necessary to solve specific problems. But a lot of this knowledge can be used in other situations; for example, a procedure to choice rolling bearing is a very general procedure for engineering design. Thus, the problems are: 1) how to represent this procedure and 2) where to store it to make it accessible to all people who have necessity to use. This means that the engineering knowledge must be stored in PLM/PDM systems; however today these systems do not provide functions and methods to do it. In our works we investigated methods to inquire in automatic way a product data base to retrieve parts to implement a function and store reusable knowledge for different MDA applications but in the same industrial context and not in general way.

Finally, we think that MDA is an important tool to improve and innovate design procedures, both in big companies and SMEs. At present, MDA is accessible also to SMEs in terms of human resources, ICT tools investments and scientific/technical competences. Moreover, EKM related to product and process is a powerful tool to disseminate best practises within company departments making people aware of company intellectual property and at the same time a way to take care of the company knowledge and know-how.

References

1. Choi JC, Kim C (2001) A compact and practical CAD/CAM system for the blanking or piercing of irregular shaped-sheet metal products for progressive working. J Mater Process Technol 110:36–46
2. Colombo G, Ferretti D, Cugini U (1989) How to represent design rules in a parametric CAD system. In: proceedings of international symposium on advanced geometric modelling for engineering applications, Berlin, Germany
3. La Rocca G, Krakers L, van Tooren MJL (2002) Development of an ICAD generative model for blended wing body aircraft design. In proceedings 9th symposium on multidisciplinary analysis and optimization, AIAA/ISSMO, Atlanta, USA
4. Craig B, Chapman MP (2001) The application of a knowledge based engineering approach to the rapid design and analysis of an automotive structure. Adv Eng Softw 32:903–912
5. Stokes M (2001) Managing engineering knowledge—MOKA: methodology for knowledge based engineering applications. Professional Engineering Publishing, London and Bury St Edmunds
6. Rosenfeld LW (1995) Solid modeling and knowledge-based engineering. In: LaCourse DE (ed) Handbook of solid modeling. McGraw-Hill Inc, New York
7. Colombo G, Cugini U, Mandorli F (1992) An example of knowledge representation in mechanical design using the object-oriented methodology. In: Proceedings of the 1992 European simulation symposium, Dresden, Germany
8. Coad P, Yourdon E (1990) Object-oriented analysis. Yourdon Press Computing Series, Prentice Hall, Englewood Cliffs NJ
9. Colombo G, Cugini U (1992) Knowledge aided design; requirements, systems and applications. Revue Internationale de CFAO e d'Infographie 7(3):293–309
10. Germani M, Mandorli F (2004) Self-configuring components approach to product variant development. AIEDAM Spec Issue: Platf Prod Dev Mass Customization 18(1):41–54
11. Raffaeli R, Germani M, Graziosi S, Mandorli F (2007) Development of a multilayer change propagation tool for modular product. In: Proceedings of the ICED 2007, 16th international conference on engineering design, Paris, France
12. Mandorli F, Rizzi C, Susca L, Cugini U (2001) How to implement feature-based applications using KBE technology. In: Proceedings of the IFIP-FEATS 2001 international conference on feature modeling and advanced design-for-the-life-cycle systems, Valenciennes, France
13. Susca L, Mandorli F, Rizzi C, Cugini U (2000) Racing car design using knowledge aided engineering. AIEDAM 14(3):235–249
14. Germani M, Mandorli F, Otto HE (2003) Product families: a cost estimation tool to support the configuration of the solution phase. In: Folkeson A, Gralen K, Norell M, Sellgren U (eds) Proceedings of the ICED 2003, 14th international conference on engineering design, Published by Design Society, Stockholm, Sweden
15. Raffaeli R, Alfaro D, Germani M, Mandorli F, Montiel E (2006) Innovative design automation technologies for corrective shoes development. In: Proceedings of the 9th international design conference DESIGN 2006, Edited by Marjanovic D, Dubrovnik, Croatia
16. Colombo G, Pugliese D and Rizzi C (2008) Developing DA applications in SMEs industrial context. In: Proceedings of IFIP-CAI 2008, Milan, Italy
17. Colombo G, Pugliese D, Saturno Spurio M (2007) About the integration between KBE and PLM. In: Proceedings of CIRP-LCE 2007, Springer Verlag, Tokyo, Japan
18. Colombo G, Facoetti G, Gabbiadini S, Rizzi C (2010) Virtual configuration of lower limb prosthesis. In: Proceedings of the ASME 2010 international mechanical engineering congress & exposition IMECE 2010, Vancouver, British Columbia, Canada
19. Colombo G, Pugliesc D (2008) Improving product design by using design automation. In: Horvath I, Rusak Z (eds) Proceedings of TMCE 2008, Izmir, Turkey
20. Eriksson H, Penker M (2000) Business modeling with UML. Wiley, New York
21. http://www.idef.com/w.idef.com. Accessed Mar 2011

22. Forster J, Arana I, Fothergill P (1996) Re-design knowledge representation with DEKLARE. In: Pierret-Golbreich C, Fensel D, Motta E, Willems M (eds) Proceedings of KEML 1996 6th workshop on knowledge engineering: methods & languages, Paris, France
23. Raffaeli R, Mengoni M, Germani M, Mandorli F (2009) An approach to support the implementation of product configuration tools. In: Proceedings of the ASME international design engineering technical conferences & computers and information in engineering conference, IDETC/CIE 2009, San Diego, California, USA
24. Colombo G, Rizzi C, Scotto M (2010) A knowledge based application to enhance fired heater design. In: Proceedings of the IDMME-Virtual Concept 2010, Bordeaux, France
25. ANSI/API STANDARD 560/ISO 13705. Petroleum, petrochemical and natural gas industries—Fired heaters for general refinery service. 4th edn, August 2007
26. ANSI/API STANDARD 530/ISO 13704. Petroleum, petrochemical and natural gas industries—Calculation of heater-tube thickness in petroleum refineries. 6th edn, Sept 2008

Chapter 5
Styling Features for Industrial Design

Marina Monti

Abstract This chapter focuses on the concept of shape features for industrial design of product models, intended as a means for conveying a specific design intent; in geometric modelling the feature-based approach aims at facilitating the design activity while enabling shape manipulations directly linked to the semantics the user has in mind. This chapter describes, from this perspective, the evolution of the tools for computer-aided industrial design that led to the modern tools and methods for the definition of styling product shape models; an overview of the main research approaches and outcomes for the identification and the exploitation of styling features is provided. The current limitations and possible future trends are presented as well.

5.1 Introduction

The problem of representing complex shapes has been addressed since early 70s with the first CAD systems and led to the definition of the current geometric modelling systems based on complex mathematical models and able to handle whatever non-canonical shape. One area where the need to model extremely complex shapes is particularly acute is that of the industrial design, which includes the specification of all those products that are somehow subject to visual judgment and appreciation. It ranges from transport vehicles to home appliances, furniture, cosmetic containers and many other objects that we deal with in our daily life. The visual appearance is a key factor for the product success; customer's choice for products providing almost the same functionalities and comparable quality is

M. Monti (✉)
Istituto per la Matematica Applicata e le Tecnologie Informatiche,
Consiglio Nazionale delle Ricerche, Via De Marini 6, 16149 Genoa, Italy
e-mail: monti@ge.imati.cnr.it

M. Bordegoni and C. Rizzi (eds.), *Innovation in Product Design,*
DOI: 10.1007/978-0-85729-775-4_5, © Springer-Verlag London Limited 2011

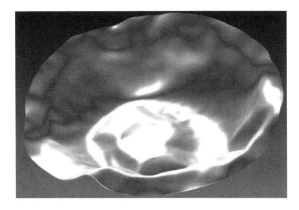

strongly based on the product appearance. Furthermore, the availability of
new-production technologies and innovative materials makes the manufacture of
very complex shapes possible, thus permitting a greater freedom of design and the
creation of products that can be considered artistic artefacts (Fig. 5.1). Moreover
such a need for freedom of expression forms is often combined with the need for
easy to use design tools able to define high-quality surfaces.

Despite the availability of computer-aided tools supporting the creation of
styling models through a wide range of modelling functionalities based on flexible
and efficient mathematical representation, designers even more required tools
more suited to their mentality that allow them to directly act on curves and
properties relevant to their design intent.

It has been argued that designers conceive a new product by making sketches of
some essential curves that are an abstraction of the product model [1, 2] (Fig. 5.2).

These curves may not only be structural lines, like profiles, but also meaningful
lines strongly affecting the product impression, which are usually referred to as
character lines. Stylists normally judge the aesthetic character of the product from
the flow of certain lines that have no explicit representation in the product model,
but can nevertheless be perceived, such as the lines originated by the reflection of
the light on the object. Thus, these lines may be considered both as properties or as
features of the model: on the one hand they reveal properties of the underlying
surface (all surface points on the lines share the same geometric property, e.g.
same angle between the normal to the surface and the light ray), on the other hand
they are feature lines with their own set of properties. We consider as *styling
features* both derived and constructive elements as long as they are connected with
the aesthetic impression of the object (in contrast to engineering features, which
modify the shape for functional or technical reasons). Styling features may also
provide information about the technical quality (e.g. surface continuity) but mainly
about the aesthetic and the emotional character of the product.

Given the importance of these lines for the final product appearance, there has
been a growing interest in the research community towards the exploitation of
these aesthetically relevant product elements for improving the design process;
research works in different fields (cognitive psychology, computer science…) have

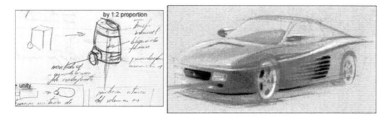

Fig. 5.2 Product sketches (Courtesy of Eiger and Pininfarina)

been carried out, aimed at exploring the possible links between shape character-istic and elicited feeling with the final objective of exploiting these possible relationships for the creation of computer-assisted tools able to support modeling functionalities, driven by the design intent.

This chapter focuses on the evolution of the research for the identification and exploitation of styling features, providing an overview of the main outcomes and also highlighting the existing drawbacks.

Section 5.2 illustrates the evolution of the design tools from the first surface modellers to the free-form feature concept. Section 5.3 provides a short overview of the research studies investigating possible relationships between form features and elicited feelings. Section 5.4 illustrates the main research results for the exploitation of styling features for the definition of computer-aided styling (CAS) tools acting on features properties directly connected to the aesthetic design intent. Section 5.5 concludes the chapter suggesting some possible research trends.

5.2 The Evolution of the CAD Tools for Styling

Modern free-form surface modelling tools were born to meet the need of designers to easily and interactively create and modify the complex shapes they have in mind. The development of free-form curves and surfaces for computer graphics began in the 60s with Paul de Casteljau and Pierre Bezier, engineers in the French automotive industry, namely at Citroen and Renault. Bezier exploited de Casteljau's research results to define a new curve representation based on Bernstein polynomials. Bezier representation allows the definition of curves which can be controlled by a set of points, named control points, connected by a control polygon that creates a sort of framework for the curve with interesting properties: the curve end points coincide with the first and the last control points and the curve is tangent to the extremes of the control polygon and it is fully included in it. The curvature can be varied by manipulating these control points. Bezier curves thus have the advantage of easily representing and manipulating aesthetic shapes and their mathematical properties allows us to reduce sensibly the number of variables necessary to handle their geometry, thus making them really suitable for the implementation of modeling algorithms.

To represent more complex shapes, in the early 70s B-spline curves have been introduced in computer-aided-geometric-design (CAGD) tools; B-splines are similar to Bezier curves, as they are controlled by the control-point polygon, but they are defined by a sequence of polynomial segments, thus allowing an higher degree of flexibility. In the late 70s, NURBS (Non-Uniform Rational B-Splines) were defined, in which the control points can be weighted and arc or conic curves can be represented without approximation.

NURBS representation have become a de facto standard for shape modelling and geometric design and it is supported by most commercial modelling systems because of its flexibility and the availability of fast and stable algorithms [3]. Research efforts during the last 20 years originated a very wide set of modelling techniques based on NURBS covering most of the functionalities needed in shape-modelling processes.

However, there are some deficiency in using NURBS-based modelling systems for styling, mainly related to the unintuitive manipulation of the control parameters; in this creative domain, the functionalities for model creation and manipulation are often perceived to restrict the way in which a shape can be modeled. The limitations are mainly due to the fact that the modelling activity is mostly based on low-level geometric elements and in order to really exploit the flexibility of the shape representation, it is necessary to fully understand the underlying mathematics, and then to concentrate too much on how to use the system itself to obtain what designers have in mind.

The need of a more intuitive modelling process, originated research activities devoted to investigating new modelling functionalities more connected with the designer intent; the form feature [4] concept appeared as a good solution to overcome these limitations. In geometric modelling, the use of form features improves the efficiency in creating a product model facilitating the exploration of alternative solutions, since a form feature can be manipulated through a limited number of parameters meaningful for the designer. In mechanical design a form feature is what makes functionally two different objects that have been initially defined by the same overall geometry; analogously in the styling domain, we may consider a form feature what makes aesthetically two different products with the same functionality (Fig. 5.3).

5.2.1 The Free-Form Feature Concept

While the features defined for mechanical CAD systems are mainly related to canonical geometric shapes, easy to classify and parameterize, and the association between shape and function is quite straightforward, in the free-form domain the definition of a free-form shaped feature class is much more complicated, as it has to be established a relationship between geometrical representation (i.e. NURBS) and a set of parameters chosen in order to make more intuitive the possible modification of the feature.

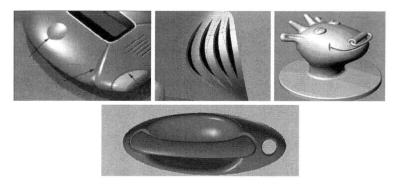

Fig. 5.3 Part of product models with free-form features (Courtesy of SAAB and Alessi)

The problem of defining free-form features as parametric primitives for the definition and the modification of product models has been addressed since the late 90: Cavendish and Marin [5] proposed a feature-based approach for the design of complex shapes, providing a method for the definition of free-form details through a surface assembly technique. Polderman and Horvath [6] proposed one of the first methods to build a geometric model by using a set of parameterized features classified in a taxonomy.

In the literature, it is possible to find several studies focusing on the user interactions with shape product through free-form features: Pernot et al. in [7] provide a comprehensive state-of-the art of the main research works undertaken to adapt the feature concept to the free-form domain, providing a critical comparisons of the different approaches. The research outcomes include various meaningful proposals for classification of free-form features and several interesting works dealing with the shape modification by features.

Among the works more focused on the styling domain, Fontana et al. in [8] proposed an interesting study of the free-form features used in the different phases of computer-assisted styling activity. Authors pointed out that in the aesthetic design, the concept of free-form features is related to two different logical phases: the product overall shape definition and the local refinements of the shape. According to these logical steps, they identified two categories of free-form features: *structural features* and *detail features* (Fig. 5.4).

Structural features have an important global aesthetic impact on the product and are used in the preliminary phase of design for defining the surfaces constituting the product; they consist of sets of evaluation lines, such as section curves, reflection lines, contours or special curvature lines, grouped according to their intrinsic meaning and handled as a unique entity, maintaining the relationships with the corresponding derived surfaces (Fig. 5.5).

Detail features are created in the second modelling phase and correspond to local shape modifications for adding aesthetic and functional details enforcing the visual effects of the so-called character lines. The study mainly focuses on the

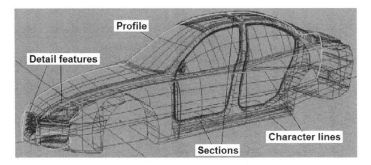

Fig. 5.4 Examples of structural and detail features in a car body model

Fig. 5.5 The lemon squeezer
by Philippe Stark in photo
and with contour lines
(Courtesy of Alessi)

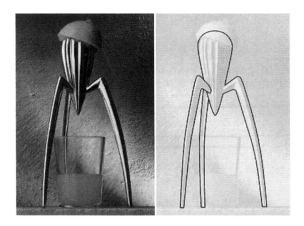

classification of detail features that is based on the topological and morphological characteristics associated to the deformations provided by the considered features.

The research in the free-form features for the CAGD domain initially seems to take into account aesthetic information only marginally, however in the research community in the early 90s the interest towards the identification of the aesthetically relevant product shape elements is growing and several studies are conducted which are aimed at capturing affective responses to product shapes while associating them with product form characteristics.

5.3 The Exploration of the Relationships Between Shape Features and Aesthetic Properties

Several investigations have been carried on for the exploration of users' emotional reactions to product form, with the main aim of providing product designers with a better understanding of consumer needs. The studies aimed to identify the relationships between product shape elements and elicited feelings are commonly using experimental approaches and are based on the analysis of the answers given by selected target groups on a collection of shapes. Kansei Engineering is probably the most known among the methodologies proposed to relate the basic shapes of a product to the emotional response elicited by the product itself [9–11].

Kansei is a Japanese word, which indicates a method for translating feelings and impressions into product parameters. Kansei engineering aims at producing a new product based on consumers' feeling and demand. Traditional Kansei engineering approaches use "items" and "categories" defined in pictographic terms to provide qualitative descriptions of the global product; surveys or experiments are conducted to evaluate the consumer's image of the product. With the identification of the design elements of the product, the relationship between Kansei (image) words and the design elements can be established. The standard procedure of a Kansei engineering experimentation is based on the semantic differential method [12]: a typical study evaluates the affective responses elicited by product shapes using a set of bipolar adjective pairs, e.g. modern–classic, customized for the particular product of interest. An experimentation based on semantic differential consists of the following basic steps: (1) Identify a set of adjective pairs expressing the aesthetics of the product under consideration. (2) Select the most suitable adjective pairs. (3) Interrogate potential users to evaluate each product design via questionnaires, yielding numerical measures for the important aesthetic factors. (4) Analyze the numerical data and utilize the results for design decision-making [13].

However, these methods present some intrinsic limitations mainly due to the subjective approach to shapes evaluation [14, 15]; even if the majority of studies are carried out with very systematic methodology, possibly involving people with different background, the valence of the obtained results cannot guarantee generally valid conclusions. In addition, normally the products considered in the experimentations were existent on the market; therefore the reaction can be noised by cultural and usage experiences of the products and not only due to the shape itself. Moreover the differences in the image words considered, and in the focus on different particular products make almost impossible a comparison among the different studies to detect discrepancies and consistencies to attain some general theory.

Despite the huge efforts, the exploitation of the possible relationship between shape features and product appreciation for the definition of computer-aided tools for styling remained limited: few works considered aspects such as brand character or emotional feeling [16–21] but are mostly peculiar to specific products and not generally applicable. Among them, Hsiao and Wang in [18] presented a method for modifying the rough model of a car according to a wished character. Their approach is based mainly on the collection of customer verbal descriptions and only relies on the car proportions such as height, tail length. Fujita et al. [19] proposed a methodology for designing, by introducing aesthetic features for interpretation of aesthetics. Van Bremen et al. in [17] provided some examples of possible, but not tested, associations between aesthetic and shape parameters without proving an effective feasibility of the mapping process. They concluded that such an association is rather difficult and is not a simple mapping, since the same aesthetic parameters can be associated with different shape parameters.

Since the difficulties in identifying stable and reliable relationships between shape characteristics and product elicited feeling prevented the definition of generic classes of styling features directly connected with the designer aesthetic intent, research efforts started focusing on the investigation and exploitation of the properties that stylists consider when they create the digital model and verbally communicate their design directives to exploit the potentiality of shape features meaningful for styling both in modelling and in evaluation phase.

5.4 Styling Features and Semantic-Based Design

In industrial design one of the main goals is developing products evoking a specific emotional character; as already mentioned the product character is perceived mainly from the *character lines*, which correspond to silhouettes, light lines, particular sections and other significant curves used to start the model design: these lines may be referred as *styling features*. Designers always face the problem of achieving a high-quality surface with a certain light effect or, more in general, characteristic curves behavior. The final results are normally obtained through a trial-and-error approach which implies a loop of successive very small and localized modifications and evaluations, as the CAS operators do not know exactly with which sequence of operations the goal is achieved, since there is not an explicit relationship between the target property and the geometric handles managed by users. At the end of 90s emerged the target-driven design approach for the design of industrial products: the basic idea is to allow designers to specify directly their targets as request of a shape generation or modification. The studies conducted in the European project FIORES [22] demonstrated that the modelling activity can be greatly improved by offering the possibility of specifying directly the desired property values to automatically obtain the corresponding surface modification [23].

A few years later, the European project FIORES-II [24] explored a wider range of aesthetic properties of styling features to be used in the target driven approach.

Designers concentrate their attention on properties that may have a local nature or a more extended behaviors, e.g. locus of points sharing the same property value, for instance light lines, or specific property combinations: they are evaluated from the surface normal, curvature, and partial derivatives. Some of the properties have been clearly identified and mathematically formalized; others still require effort to be fully understood since their most natural mathematical formalization does not fully reflect the stylists' interpretation.

5.4.1 Light Lines and Curvature Behaviour Properties

Light lines are the most well-known among the styling curves and the first for which some results have been obtained with regard to the possibility of working

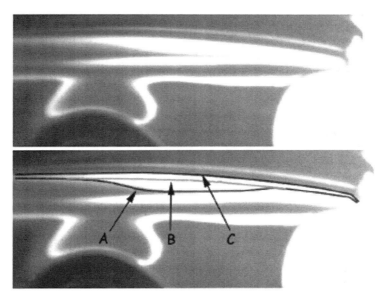

Fig. 5.6 Simulated reflection line modification on a car side panel

with them [25–28] (Fig. 5.6). Light lines include reflection lines, isophotes and highlight lines; they are used as intuitive means for detecting surface irregularities, due to the immediate common understanding of the light reflections [3, 29].

As light lines provide information on the surface quality being linked to the surface curvature, at the same time their quality and the quality of the character and constructive lines is very important. It has been recognized that stylists are more sensitive to curvature characteristics of lines and surfaces than laypeople. From practical experiments it has been seen that while laypeople identify the most important points in the curves as mainly those corresponding to the inflections and extension extremes (i.e. the highest and lowest or the most right and left points), on the contrary stylists are more sensitive to inflections and curvature extremes [30]. In fact, curvature extremes normally correspond to the junctions of the segments having monotone curvature in which stylists conceptually divide a curve. Moreover, it has long been recognized that the number of extremes in the curvature distribution, and hence the number of separate segments with monotone curvature should be the minimum required to meet the aesthetic intent of the designer: these considerations motivated the activities done to create and modify curves and surfaces according to a certain curvature behaviour [31–34].

According to the target-driven approach, a few years later as results of the European project T'nD, Bosinco et al. [35] presented sweep techniques under constraints to directly generate a shape having a highlight distribution as close as possible to a specified one.

5.4.2 Aesthetic Properties of Styling Features

At a first glance, many styling properties seem to depend directly on curvature properties, but although tests with curvature-dependent similarity functions have given reasonable results, they do not always fit the designers' thinking. To identify an appropriate characterization of shapes in terms of styling features and to understand which features properties stylists consider, the design activities carried out in different industrial design fields have been deeply analysed within the FIORES-II project [24]. Among the other results, the project showed that different languages are adopted in different phases of the product-design process (marketing, concept design and CAS modelling). In particular, during the creation and modification of the product model, stylists express detailed directives when they work with surfacers at the definition of the 3-D digital model using a limited number of terms strictly linked to shape properties: with these terms they provide instructions for the modification of the product shape (e.g. "making a curve a bit more *accelerated*", or "decreasing curve *tension*").

In particular, the study focused on how stylists talk about styling properties for communicating their ideas. It emerged that they use different terms when speaking with marketing people or when working with CAS operators at the definition of the 3-D digital model. In the former, the terms used have an emotional value (e.g. aggressive, elegant…) while in the latter they provide an indication on which geometric elements and related shape properties have to be changed to obtain the desired effect. Normally the styling directives expressed in these terms are executed by CAS operators, which are able to translate them into the expected results throughout sequences of modelling operations, not directly linked with the target properties. This is possible only thanks to a great skill both in modelling and in the adopted tools, but often requires a time-consuming trial-and error loop. Therefore, there are clear advantages derived from the exploitation of the knowledge implicit in these terms. The terms referring to these properties represent a first link between low-level CAGD descriptions and the high-level character of a product. The list of the identified terms is neither complete nor do stylists use all the identified concepts. Nevertheless, all stylists, designers and model makers involved in the FIORES-II project agreed that the list was reasonable, even though they came from four different European countries and worked in different domains, like automotive and consumer appliance industries. The following terms were recognized as the most used to communicate design intent:

- Radius/Blending
- Convex/Concave
- Tension
- Straight/Flat
- Hollow
- Lead-in
- Soft/Sharp
- S-Shaped

- Crown
- Hard/Crude
- Acceleration

Some of the aesthetic property terms have been translated in modelling and measuring operators directly acting on the geometry to obtain the desired shape results. To implement such algorithms the terms identified have been formally described and quantified; for each considered styling property the following steps were necessary:

- The definition of its meaning from the designer's point of view (i.e. what shape is the designer expecting when the property value changes for the considered entity);
- The identification of the geometric properties that are affected by the styling property;
- The setup and evaluation of a measure function;
- The specification of the mathematical function producing the expected shape modification and its related domain of application.

In this context there are several difficulties in fulfilling the above tasks, mainly related to getting a full comprehension on how stylists perceive shape and then to translate this into mathematical formalism. Even if some of the terms used have a direct mathematical counterpart, the meaning is not exactly the same; for example not all the curves in which the second-order derivative increases are necessarily perceived as accelerating curves. Moreover, different shapes may be perceived as having the same property value. This means that several characteristics/variables contribute to a single property, thus requiring a further level of interpretation to give a formal description of both the property and its measure. In [36, 37] authors proposed a measure for each term that constitutes the basis for the implementation of the related operator in the software prototype (Fig. 5.7).

In the last decade the availability of powerful and flexible knowledge technologies has brought big benefits to the CAD paradigm, allowing for the development of product modelling functionalities able to incorporate prior knowledge into the product model, thus integrating all the information produced throughout different design phases. To make possible the exploitation of this aspect for industrial design process, research works focused on the formalization and management of the aesthetic knowledge. In this perspective the styling features become a fundamental element in the formalization of the semantics embedded in the product concept model.

5.4.3 Use of Styling Features in the Semantic-Based Design

In the latest years many research approaches aimed at supporting concept design proposed a semantic-based approach based on form-features meaningful for styling. Catalano et al. in [38] proposed a framework supporting the semantic

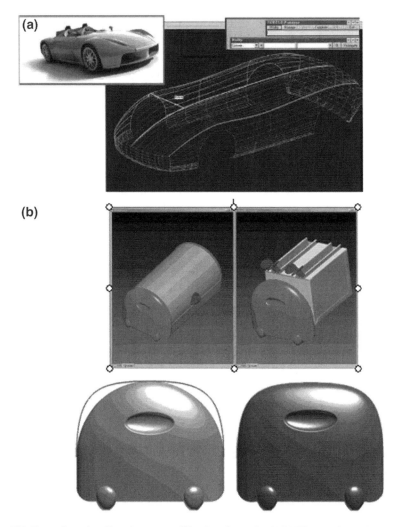

Fig. 5.7 Examples of styling feature modification from the FIORES-II project (Courtesy of Pininfarina (**a**) and Alessi (**b**))

annotation of digital model of cars; authors captured and structured into an ontology the styling semantics embedded in the 3-D models and 2-D drawings representing automotive products. It includes a taxonomy of the styling features in the automotive field, the relations between them and the related aesthetic properties. A descriptive model of the automobile styling is proposed also by Zhao et al. in [39].

In the context of a semantic-based approach for curve modelling, the AIM@SHAPE project [40] proposed a conceptualization of a multi-layered architecture according to which shape models are organized at three different levels of increasing information abstraction: geometric, structural and semantic layer. The geometric level is the representation of the shape data, in which different types of geometric models can be used (e.g. NURBS and 3D mesh). At the structural level elements are grouped and associated to abstracting elements according to some shape or topological characteristic (e.g. curvature evolution and connected object components). At the semantic level the association between shape model parts and their associated semantics is made explicit, through annotation of shapes according to the specific application domain.

At the lower geometric layer, the geometric representation enables direct manipulation of the shape, through the interaction with its control variables or by imposing specific constraints on the geometric parameters defining the shape.

In the semantic level it is possible to exploit all the available knowledge to create semantic-driven operators which act on styling features and their aesthetic properties, strictly connected to the stylist's point of view, thus allowing a much more simple and natural mode of interaction.

The aim of the structural layer is to fill the gap between the two other layers and to be the interface for sharing information and data with the other actors of the product-design activity, allowing an independence from possible different geometric construction methods and representations. For this reason, this intermediate level should be easily and automatically obtainable regardless of how the object is represented and built. The structural level is seen as a neutral environment in which the curves are classified and treated through the intrinsic information related to the curvature.

Conforming to this structured approach, Cheutet et al. in [41] in order to capture and structure the semantics embedded in the first sketches representing the product, proposed an ontology to guide more easily the generation and manipulation of curves. This ontology includes a taxonomy of the aesthetic curves in the automotive field and a curve manipulation setting based on a shape grammar, creating explicit relationships between the two contexts. Shape grammars provide a description of a shape through a concise and repeatable language, which acts only on the shape geometry and not on the underlying digital representation. Moreover, comparisons between shapes can be made through their grammatical characterizations. In this approach shape grammar is used for internal description and is not explicit for users; it can be considered as a neutral language between the user's knowledge and the geometric description of the shape needed by the system. This work demonstrated that styling features can be exploited for a structured representation of the semantics embedded in the product sketches in order to make possible expressing curve manipulation process in terms of shape features and operators easily understandable by stylists.

5.5 Conclusions and Perspectives

The problem of understanding relationships between product shape and consumers' appreciation is very complex and huge; it is affected by many different aspects including cultural background and environment as well as of product typology; as a consequence, the identification and parameterization of free-form features class, independent from product typology and context, directly linked with the aesthetic intent is hard to be achieved.

Anyhow, some more stable relations for well-structured product categories have been identified and have been proven to be very helpful for systems devoted to the conceptual design phase in which the general shape of the product is defined, as in this phase it is sufficient to handle only few parameter dependencies, and it is not necessary to take into account additional elements, as needed in detailed design.

The research outcomes obtained so far demonstrated that it is possible to define aesthetic-wise tools based on styling features and their properties, which can allow designers to reach their target through an optimized product design cycle, whatever the kind of product (automobiles, consumer appliances, etc.); tools able to deal with shape semantics may significantly improve computer-assisted design process, minimizing the model modification and evaluation loop. Some styling features and properties can be formally described and measured, and some of the measures proposed so far seem suitable to be used in interactive working procedures [24], with significant results in terms of working time reduction. In principle measured styling properties may also be used for shape comparisons: however in this context the wish of finding a general similarity measure for arbitrary curves has to be regarded as impossible. The similarity assessment based on local styling properties may give good results if applied to very similar objects presenting small shape variations and the same number of characteristic lines. Therefore, it seems acceptable for the evaluation of the impact of small changes, but it appears too limited for a general cataloguing and retrieval.

Even if in general the styling property measures do not discriminate univocally the curves, they can be used for the development of tools for curve modification. In this perspective the exploitation of shape semantics for the definition of tools supporting the creation of new product shapes which more possibly will satisfy users' expectation, and at the same time, supporting the fast modification of the shape, when the desired response is not achieved, has been proved to be a fruitful research direction. The ability to capture the design intent and the mathematical definitions of aesthetic properties of shapes can constitute a starting point for further research on shape semantics in the industrial design. However it has to be underlined that all the findings in this field require a very extensive and robust experimentation and validation including different actors of the design process (i.e., designer, CAD operator, consumer and psychologist), and research outcomes in this field can be achieved only thanks to an integrated multi-disciplinary approach.

An interesting and promising approach that could benefit of the research outcomes achieved for styling features and that might also bring new and interesting insights in the study of the relationships between shape features and emotion can derive from the exploitation of the enactive technologies [42]. In the European project SATIN [43, 44] a proof-of-concept of a multi-modal system was developed, based on a haptic strip which uses position sensors as an input to the reference geometry properties and render them as metaphoric sounds; the project showed how multi-modal interaction can sometimes be used to augment the user perception by conveying information through the virtual environment that somehow is not perceived in the real world.

A scenario in which users' creativity may find diverse and direct ways of expression and representation and in which user reaction to shapes at different sensorial levels may be captured, would allow to address the identification of styling features classes from a different and much more comprehensive perspective.

Acknowledgment The author would thank in particular Prof. Bianca Falcidieno and Dr. Franca Giannini for their special contribution to the topics addressed in this chapter.

References

1. Tovey M (1992) Intuitive and objective processes in automotive design. Des Stud 13:22–41
2. Catalano CE, Falcidieno B, Giannini F, Monti M (2002) A survey of computer-aided modelling tools for aesthetic design. J Comput Inf Sci Eng 2:11–20
3. Farin GE (1996) Curves and surfaces for computer-aided geometric design: a practical guide. Academic Press, San Diego
4. Shah JJ, Mantyla M (1995) Parametric and feature-based CAD/CAM. Wiley, New York
5. Cavendish JC, Marin SP (1992) A procedural feature-based approach for designing functional surfaces. In: Hagen H (ed) Topics in surface modelling. SIAM Geometric Design Publications, Philadelphia
6. Poldermann B, Horvàth I (1996) Surface design based on parameterized surface features. In: Horvath I, Varadi K (ed) Proceedings of international symposium on tools and method for concurrent engineering, Budapest
7. Pernot JP, Falcidieno B, Giannini F, Leon JC (2008) Incorporating free-form features in aesthetic and engineering product design: state-of-the-art report. Comput Ind 59:626–637
8. Fontana M, Giannini F, Meirana M (1999) A free form features taxonomy. Comput Gr Forum 18:107–118
9. Nagamachi M (1989) Kansei engineering. Kaibundo Publishing Co, Tokyo
10. Nagamachi M (1995) Kansei engineering: a new ergonomic consumer oriented technology for product development. Int J Ind Ergon 15:3–11
11. Tanoue C, Ishizaka K, Nagamachi M (1997) Kansei engineering: a study on perception of vehicle interior image. Int J Ind Ergon 19:115–128
12. Osgood CE, Suci GJ, Tannenbaum PH (1957) The measurement of meaning. University of Illinois Press, Urbana and Chicago
13. Himmelfarb S (1993) The measurement of attitudes. In: Eagly AH, Chaiken S (eds) Psychology of attitudes. Thomson/Wadsworth, Belmont
14. Ishihara S, Ishihara K, Nagamachi M, Matsubara Y (1997) An analysis of Kansei structure on shoes using self-organizing neural networks. Int J Ind Ergon 19:93–104
15. Hsiao KA, Chen L (2006) Fundamental dimensions of affective responses to product shapes. Int J Ind Ergon 36:553–654
16. van Bremen EJJ, Sudijone S, Horvath I (1999) A contribution to finding the relationship between shape characteristics and aesthetic appreciation of selected products. In: Lindemann U, Birkhofer H, Meerkamm H, Vajna S (ed) Proceedings international conference on engineering design ICED 99, The design society, Munich

17. van Bremen EJJ, Knoop WG, Horvath I, Vergeest JSM, Pham B (1998) Developing a methodology for design for aesthetics based on analogy of communication. In: Proceedings 1998 ASME design engineering technical conferences, Atlanta
18. Hsiao SW, Wang HP (1998) Applying the semantic transformation method to product design. Des Stud 19:309–330
19. Fujita K, Nakayama T, Akagi S (1999) Aesthetics function and geometry with feature-based modelling and constraint management. In: Lindemann U, Birkhofer H, Meerkamm H, Vajna S (ed) Proceedings international conference on engineering design ICED 99, The design society, Munich
20. Chen K, Owen CL (1998) A study of computer supported formal design. Des Stud 19:331–358
21. Smyth SN, Wallace DR (2000) Towards the synthesis of aesthetic product form. In: Proceedings of DETC'00 ASME 2000 design engineering technical conferences and computers and information in engineering conference, Baltimore
22. FIORES Formalization and integration of an optimized reverse engineering styling workflow, Brite Euram-Project n. BE96-3579. http://www.fiores.com. Accessed 18 February 2011
23. Bosinco P, Durand G, Goussard J, Lieutier A, Massabo A (1998) Complex shape modifications. In: Batoz JL, Chedmail P, Cognet G, Fortin C (ed) Proceedings 2nd IDMME Conference, Kluwer Academinc Publisher, Dordrecht
24. FIORES II, Character preservation and modelling in aesthetic and engineering design, GROWTH Project n. G1RD-CT-2000-00037, http://www.fiores.com. Accessed 18 February 2011
25. Chen Y, Beier K, Papageorgiou D (1997) Direct highlight line modification on Nurbs surfaces. Comput-Aided Geom Des 14:583–601
26. Zhang C, Cheng F (1998) Removing local irregularities of NURBS surfaces by modifying highlight lines. Comput-Aided Des 30:923–930
27. Andersson R (1996) Surface design based on brightness intensity or isophotes-theory and practice. In: Hoschek J, Kaklis P (eds) Advanced course on FAIRSHAPE, B. G. Teubner gmbh, Stuttgart
28. Loos J, Greiner G, Seidel H (1999) Modeling of surfaces with fair reflection line pattern. In: Proceedings of shape modeling and applications, IEEE Computer Society Press, Los Alamitos
29. Hagen H, Hahmann S, Schreiber T, Nakajima Y, Wordenweber B, Hollemann-Grundstedt P (1992) Surface interrogation algorithms. J IEEE Comp Gr Appl 12:53–60
30. Podhel G (2001) How designers see curvature properties. University of Kaiserlautern, RKK Internal Report RIR-2001–2002
31. Burchard HG, Ayers JA, Frey WH, Sapidis NS (1994) Approximation with Aesthetic Constraints. In: Sapidis NS (ed) Designing fair curves and surfaces, society for industrial and applied mathematics, Philadelphia
32. Higashi M, Tsutamori H, Hosaka M (1996) Generation of smooth surfaces by controlling curvature variation. J Comput Gr Forum 15:187–196
33. Anderson RE (1993) Surface with prescribed curvature. Int J Comput-Aided Geo Des 10:431–452
34. Frey WH, Field DA (2000) Designing Bezier conic segments with monotone curvature. Comput-Aided Geo Des 17:457–483
35. Bosinco P, Massabo A, Rejneri N (2006) Touch and design. In: Proceedings 7th international conference on computer-aided industrial design and conceptual design (CAID & CD'2006), IEEE press, Hangzhou
36. Podehl G (2002) Terms and measures for styling properties. In: Marjanovic D (ed) Proceedings of the 7th international design conference, The Design Society, Dubrovnik
37. Giannini F, Monti M, Podhel G (2006) Aesthetic-driven tools for industrial design. J Eng Des 17:193–215
38. Catalano CE, Giannini F, Monti M, Ucelli G (2007) A framework for the automatic annotation of car aesthetics. Artif Intell Eng Des Anal Manuf 21:73–90

39. Zhao D, Zhao J, Tan H (2009) A feature-line-based descriptive model of automobile styling and application in auto-design. In: Proceedings international conference IASDR2009, Korean Society of Design Science, Seoul
40. AIM@SHAPE (Advanced and Innovative Models And Tools for the development of Semantic-based systems for Handling, Acquiring, and Processing knowledge Embedded in multidimensional digital objects), European Commission Contract IST 506766, http://www.aimatshape.net. Accessed 18 February 2011
41. Cheutet V, Catalano C, Giannini F, Monti M, Falcidieno B, Leon J (2007) Semantic-based operators to support car sketching. J Eng Des 18:395–411
42. www.enactivenetwork.org. Accessed 18 February 2011
43. Bordegoni M, Cugini U (2008) The role of haptic technology in the development of aesthetic driven products. Special issue on haptics, tactile and multimodal interfaces. J Comput Inf Sci Eng 8:1–10
44. Bordegoni M, Cugini U (2009) Multimodal perception-action interaction. In: Proceedings IDETC/CIE ASME 2009 conference, ASME, San Diego

Chapter 6
The Evolving Role of Computer-Aided Engineering: A Case Study in the Aeronautical Structural Design

Pietro Cervellera

Abstract This chapter discusses some aspects of how the evolution of computer-aided engineering has affected product design practise. As a case study we describe the main challenges of aeronautical structural design and how the role of Computer-Aided Engineering—CAE evolves to deliver fundamental process improvements. It is outlined the double function of simulation: early in the initial design phases, it supports engineering and decision-making, whereas in later phases it helps in validating the design with respect to specifications. The validation function is widely practised in the industry and this is where most investment in simulation is made. While its importance is recognized, the engineering function is still underestimated in terms of technology availability, resources and investment. Zooming-in on the technology aspect, two industrial applications of Finite Element-based structural optimization are presented, which illustrates the important impact of such technology on the engineering function. These examples allow us to lead the discussion further on how to close the gap between "state-of-the-art" technology and its exploitation in "state-of-the-art" processes.

6.1 Design Challenges in Aeronautical Structural Design

Aircraft structural engineering faces several challenges common to other areas of highly-engineered product design. These can be ultimately summarized in the balancing of design performance, development time and costs.

Design performance in aeronautical structural engineering can be seen as the ability of the structure to meet specifications in terms of function and safety. The second one is of course of primary interest: it is not just that the integrity of every

P. Cervellera (✉)
Altair Engineering, Munich, Germany
e-mail: cervellera@altair.de

M. Bordegoni and C. Rizzi (eds.), *Innovation in Product Design*,
DOI: 10.1007/978-0-85729-775-4_6, © Springer-Verlag London Limited 2011

structural component must be guaranteed within the full spectrum of situations to be met by the aircraft during its life cycle, but it must also show enough reserve to enable it to cope with unseen eventualities as well. This aspect is so critical that aviation authorities have established formal certification processes in order to force manufacturers to provide evidence that structures meet requirements.

On the other hand a major indicator of structural design "quality" is weight. It can be demonstrated by solid mathematics, but is also common sense, that, the lighter the structure is, the better the aircraft will perform in flight. In particular for commercial airliners the relation between weight and commercial success is evident. At the end of 2003, Airbus announced [1] that the A380 was going to exceed the weight target by 9 tons, which means that the maximum take-off weight (MTOW) was of 569 tons instead of 560 tons, i.e. + 1.6%. This weight penalty would have led to major consequences: the aircraft's performance was reduced (reduced flight range, + 120 m to the take-off run), commercial performance was negatively affected (+ 1% in operating costs) and there was an increase in the development costs since local structural reinforcements were required.

The complexity of the aircraft-product, and the variety of design requirements results in development programs spanning several years, and costs in the order of billions. Just to provide some figures, the Airbus A380 program was launched on 19 December 2000 and the first aircraft was delivered to Singapore Airlines on 17 October 2007, nearly seven years later [2]. The Airbus A350 development costs are currently projected at US$15.2 billion [3].

As initially stated, the ultimate challenge is to balance design performance, development time and costs from both planning and operational perspectives, thus meeting the planned goals. An example is the trade-off between the increased design complexity and manufacturing costs of composite materials and their increased structural performance, in particular their lower weight. Both Airbus and Boeing in their most recent developments, the A350 (53% composites) and the B787 (50% composites), opted to increase their adoption of composites, reducing aircraft weight, resulting in lower specific fuel consumption and therefore operating costs.

The idea of "right first time, first right time" perfectly applies to aeronautical structural design: with such long and costly development cycles, delays due to additional design iterations can have catastrophic effects on time-to-market. For example, in June 2009 Boeing announced 6 months delay on the B787 maiden flight due to a "need to reinforce an area within the side-of-body section of the aircraft", a problem that arose only in "scheduled tests on the full-scale static test airplane" [4].

6.2 The Role of CAE in the Aerostructures Design Process

Due to the complexity of the design process as well as the challenges and costs related to building and testing prototypes, the aerospace industry has been a leader in adopting computer-based analysis. In particular, the first general purpose

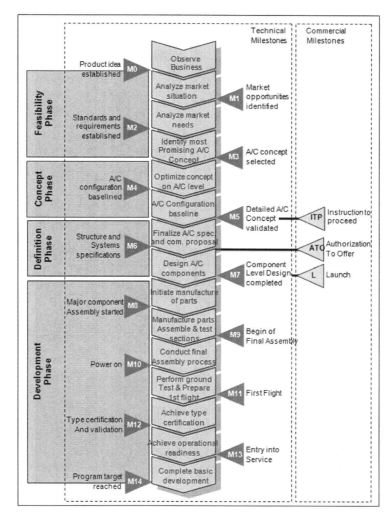

Fig. 6.1 Aircraft development process and milestones (adapted from [9])

Finite Element Analysis (FEA) code NASTRAN (NASa Structural Analysis), was developed under a NASA contract as early as the 1960s [5]. Beyond FEA, both classical analytical formulas are present in several industry-standard handbooks [6–8] and are widely used.

In current industrial practice, simulation and analysis are used with different methods, tools and goals depending on the design phase (see Fig. 6.1). In the initial concept and pre-design phases (before milestone M6) simulation is used to support engineering and decision-making, while in detailed design (post molestone M6) it has a design validation function.

6.2.1 Simulation for Engineering (Simulation-Driven Engineering)

As in any development process of complex systems, aircraft design faces the typical paradox shown in Fig. 6.2. The main design decisions are taken in the early design stages when very limited product knowledge is available. These decisions play a fundamental role in freezing the design architecture and the costs. For instance, it is estimated that 65% of aircraft life cycle costs are locked at the end of conceptual design, and 85% at the end of the preliminary design [10].

Several efforts have been made to provide engineers with better tools and processes, especially for these early design phases. As shown in Fig. 6.3, the overall goal is the increase of knowledge and freedom upfront in order to minimize the time spent in the detailed phase. This would eventually result in a drastic reduction in the overall development time and costs.

In order to exploit the potential of virtual engineering for these purposes, in 2004, the European Aerospace industry launched a 4-year research project called VIVACE and intended, among other goals, to achieve a 5% cost reduction in aircraft development, as well as a 5% reduction of the development phase of a new aircraft design [11]. As represented in Fig. 6.4, an improvement and increase in investment through virtual engineering in the pre-M6 phase contributes to a reduction of post-M6 costs.

The main purposes of simulation in the pre-M6 phase are:

- Support architecture and material selection through quantitative estimation of structural performance;
- Feasibility studies, what-if studies;
- Weight estimation with increasing accuracy;
- Rapid structural sizing of complete aircraft or major components.

The characteristics of tools and methods designed to meet these goals can be summarized as follows:

- *Being generative*: the tools must suggest a design configuration and help to rank variants;
- *Being rapid*: due to the number of variants considered (the more the better), the tight scheduling, and the limited availability of staff, the tools must deliver results in a short timescale;
- *Being comprehensive*: ideally all necessary performances should be evaluated in order to minimize re-design risk;
- *Being moderately accurate*: as design knowledge is limited (e.g. loads may not be fully available or only estimated) there is no need for high accuracy.

Since the traditional geometry creation in CAD and subsequent integration into the DMU is typically a long process, simulation often starts from rough models created from scratch. These Global Finite Element Models (GFEM) are used for preliminary structural assessment as well as subsequent evaluation of internal

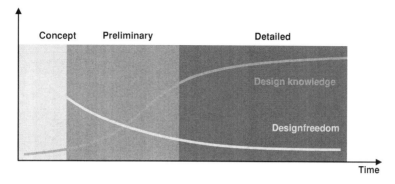

Fig. 6.2 Design paradox

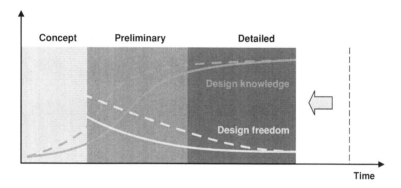

Fig. 6.3 Enhanced design process

loads for more refined local analysis in a typical multi-scale process (Fig. 6.5). As details are then added to the design, the GFEM is updated accordingly.

The GFEM analysis is purely a rough performance evaluator: to support decision-making and architecture selection, a set of different models are typically created (fuselage barrels with different frame/stringer pitch, for example) and the FE analysis provides stress/stiffness/weight estimations for comparison. The sizing of the aircraft is in most cases performed by hand, and early weight estimations are based on past experience with similar designs and statistical approaches.

A wide range of complex numerical optimization[1] programs have also been proposed to perform automatic sizing of each generated configuration. This last approach is of course much more appealing (and complex) because it outputs a set of fully-sized designs meeting all specified mechanical requirements with minimum weight. For example a research work reported in [13] shows how commercial

---

[1] In this chapter we refer to *numerical optimization* as the coupling of numerical optimization techniques with structural performance evaluation by means of FEA and analytical formulas; see for example [12].

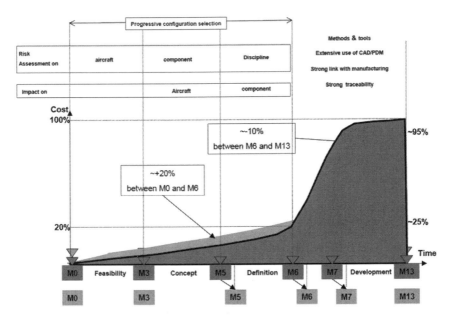

Fig. 6.4 VIVACE goals, from [11]

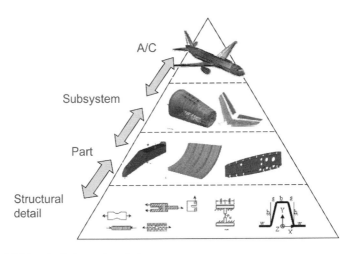

Fig. 6.5 Multi-scale design and analysis

FEA and optimization packages have been linked to in-house skills tools to size the Do728 wing at Fairchild Dornier, and another work presented in [14] describes how the solver/optimizer Lagrange developed in-house was used to size the A400 M rear fuselage at EADS-Military Aircraft.

Not all manufacturers have developed or integrated such state-of-the-art sizing systems, which require considerable expertise and investment. Referring again to

Fig. 6.2, these more traditional approaches deliver less product knowledge upfront. In other words, less knowledge is available with less quantitative accuracy.

6.2.2 Simulation for Design Validation

Figure 6.4 shows how costs explode between M6 and M7 as the aircraft enters the detailed design phase. At this time, the design is being finalized and each structural component is validated. The validation consists of proving that the component has sufficient reserve to sustain loads, and a stress report is generated for certification purposes. This process has to be carried out for each structural component at each level of detail; one reason why costs increase so dramatically is that the number of components is very high. The main characteristics of methods and tools used for design validation are the following ones:

- High accuracy: the goal of design validation accuracy in prediction is of course the key aspect;
- High level of standardization and harmonization: in modern aircraft design methods and tools are provided by the manufacturer to the complete supply chain to assure quality and reliability;
- High level of automation: as the number of components is very high and the number of load cases typically in the order of thousands, the verification process must be as automated as possible in order to reduce validation time and minimize human errors, but also to minimize staff and costs.

As mentioned, the two main methods used for structural validation are analytical methods, often implemented in several in-house codes, and detailed finite element analysis (using high-fidelity detailed FE models, named DFEM).

Analytical methods have been used for decades before FEM becomes mainstream, and each manufacturer typically has extensive knowledge of closed-form and empirical formulas with which to predict the behavior of most common structural elements. These formulas have been validated with physical tests in decades of design experience and are highly trusted. Beyond the industry-standard handbooks already mentioned, each manufacturer has its own design manual.

The implementation of these formulas in in-house software codes is necessary to provide the required ease-of-use, standardization and automation. Early implementations were fully in-house, like Festigkeit 2000, a tool developed by EADS-Military Air System [15], while the most recents are based on customized commercial platforms, like the Airbus sizing/analysis tool ISAMI (Improved Structure Analysis Multidisciplinary Integration) currently used for the A350 development [16].

The main limitation of analytical methods is the lack of flexibility: they work very well for standard structural configurations such as classical stiffened panels but there may be a lack of available methods for non-standard geometries or materials. If a method for a new configuration is not available, the development process can be very long, involving theoretical investigations and testing.

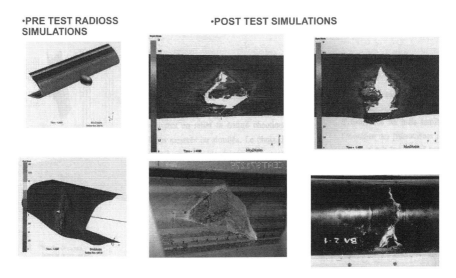

•PRE TEST RADIOSS SIMULATIONS   •POST TEST SIMULATIONS

Fig. 6.6 Bird strike analysis and test comparison [17]

For example, the buckling formula for rectangular plates has been known and applied for many years (see for example [18] from 1936) but for triangular plates it was published only in 2001 [19]. Also for innovative materials, such as composites, new methods must be developed and validated. If, for carbon fiber laminates knowledge is widely available, this may not be true for more recent materials such as Glass-Reinforced Fiber Metal Laminate (GLARE, [20]).

Detailed finite element models are increasingly being used not only for validation of non-standard structural components but also for non-standard requirements, including disciplines for which no analytical/empirical method is available (e.g. vulnerability, see pre- and post-test validation of bird strike analysis on wing leading edge in Fig. 6.6).

In general, stress analysis by means of FEA is much more flexible then analytical methods and is not limited to standard geometrical configuration. Also, knowledge on FEA methods for composite materials is rapidly increasing prediction accuracy. The main limitations in the broad adoption of FEA are also driven by the time required to create high-quality DFEM. These models must then be updated any time the design is changed and a new analysis loop is (as already noted, with perhaps thousands of load cases) carried out. Even if this may seem prohibitive, state-of-the-art pre- and post-processing tools with appropriate customization and modern hardware resources could make this possible.

In this sense the aerospace industry is following the example of the automotive industry in which structural validation relies completely on FEA. The gap between the two is due to the fact that the main design requirements for a car (stress/stiffness analysis of difficult geometry, durability and crash analysis) cannot be analyzed analytically with the desired accuracy. This is of course the result of a

Example: Bracket justification process

• Prototype process to
  automate justification
  of cable halters

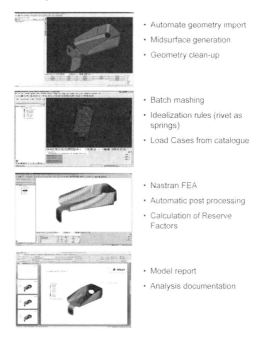

• Automate geometry import
• Midsurface generation
• Geometry clean-up

• Batch mashing
• Idealization rules (rivet as
  springs)
• Load Cases from catalogue

• Nastran FEA
• Automatic post processing
• Calculation of Reserve
  Factors

• Model report
• Analysis documentation

Fig. 6.7 Automated justification process of cable halter

constant evolution in which, driven by the demands of the market, the automotive industry has invested for decades and has partnered with software vendors to improve tools and processes for their specific needs. As an example, it has recently been demonstrated how, using state-of-the-art technology, a full car crash analysis loop can be performed in less than 24 h [21]. This highly-automated process includes geometry meshing, full car model assembly, dummy positioning, load case definition, analysis on parallel high-performance computing, result post-processing and documentation.

Even if such a level of automation is currently not possible for a complete aircraft, it is applicable for a variety of components. As an example, Fig. 6.7 (courtesy of Altair Engineering[2]) shows an automated process, derived from the automotive process discussed above, for the justification of cable halters. The simplicity of the component and its justification is evident. Nevertheless a high number of such brackets is present on a commercial aircraft (in the order of tens of thousands) and a stress report for certification is required for each one. Such a

---

[2] The software package HyperWorks has been used for this application. In particular HyperMesh/View and the Batchmesher have been used for automated pre- and post-processing, and Radioss has been used for FEA.

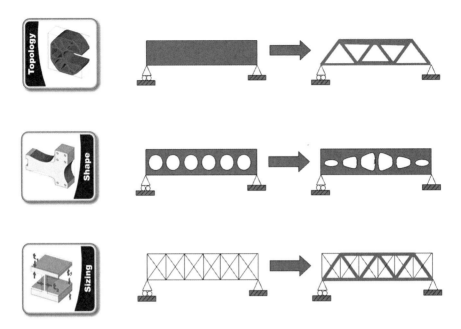

Fig. 6.8 Structural optimization disciplines from [22]

process could therefore be of great help to speed-up the complete validation process.

6.3 Optimization-Assisted Design

In Sect. 6.2.1, we have described how numerical optimization coupled with FEA is successfully used by some manufacturers not only for rapid sizing purposes, but also how implementing such complex in-house sizing systems requires considerable investment and know-how.

On the other hand, in the last decade, several commercial packages have reached such a maturity as to be successfully employed by FE analysts with no particular numerical optimization or programing skills. These packages are implemented as integrated solutions for FEA and optimization or purely as an optimizer to be coupled with external FEA solvers.[3]

Moreover these packages exploit not only sizing optimization but also other disciplines such as topology and shape optimization. As shown in Fig. 6.8, topology, size and shape optimization can assist a design task in many ways,

---

[3] Some examples of commercial packages available on the market are: OptiStruct and HyperStudy from Altair Engineering, Tosca from FE-Design, Genesis from VR&D, Nastran Sol200 from MSC.Software, BOSSquattro from Samtech.

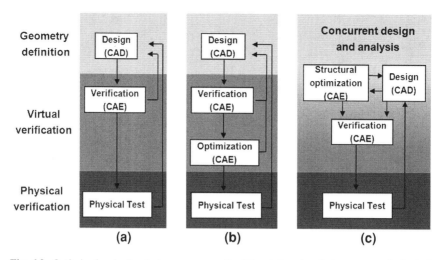

Fig. 6.9 Optimization in the design process: **a** Traditional iterative design process; **b** Optimization to improve existing design; **c** optimization-driven design process

beginning from the definition of material layout in the concept phase and ending with the final sizing and shape tuning [22].

Several recent publications discuss how optimization disciplines can be exploited to help structural design and how they can be integrated in well-established processes. As an example, here are some industrial applications: [23] discusses its application on launcher sections, [24] is relevant on aircraft flaps, [25] deals with aircraft floor structure and [26] deals with different composite sub-systems.

This leads to the idea of a process in which optimization assists or drives structural design. As schematically represented in Fig. 6.9, optimization helps to reduce classical design-analysis iterations and encourages a concurrent approach.

The following sections present two examples that illustrate how structural optimization has been used to assist the design of a simple isolated machined bracket, and to assist the concept definition of a rear fuselage section.

6.3.1 Fairchild Dornier 728 Support Arm

In this first example, structural optimization[4] has been used at Eurocopter to redesign a machined bracket. The task was to redesign this component

---

[4] The support arm optimization was performed by Hans Gruber, consultant at Altair Engineering, for Eurocopter. The software package HyperWorks 8.0 was used [22]. In particular OptiStruct was used for FEA and optimization while HyperMesh/View for pre-/post-processing and morphing.

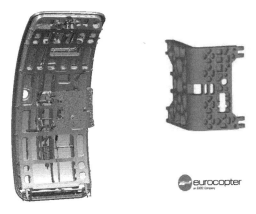

Fig. 6.10 FairchildDornier
728 Door Support
Arm—original design

maintaining the stiffness of the original design and reducing the weight by
feasible strength. As shown in Fig. 6.10, the support arm connects the door to
the fuselage, supporting its weight during the opening motion. The original
construction is an integral aluminum plate made lighter with machined pockets.
The design requirements were: (a) stiffness for three selected load cases
(blockage of door mechanism, emergency opening, a hit of the damper mechanism on the door), (b) strength.

Topology optimization was used first to produce a more efficient material
layout with optimal lightening pockets and stiffening ribs. Shape optimization was
subsequently used for fine-tuning.

Let us consider the first step. A definition of a design space is straightforward:
in this case it is imposed by packaging considerations. Furthermore, the decision
taken was about to maintain unchanged the original interface points with the rest
of the system parts, i.e. the hinges.

The optimization constraints are also straightforward: in order to maintain the
original stiffness, node displacements in the original model were measured for the
three given load cases and constrained in the optimization model. The stress
constraints for topology optimization were not available at the time. The next
requirement was a manufacturing method. Again, it was decided to use a machine-
milled aluminum plate. This aspect was introduced with a definition of draw
directions of manufacturing constraints.

The result is shown as an iso-surface of the elements' density 0.3 in Fig. 6.11.b.
Topology optimization has provided very clear material layout. The use of manufacturing constraints has minimized the effort to re-interpret it. It should be noted
that the new design substantially differs from the original one and is less intuitive.
The output of topology optimization is a rough FE model, often seen as a "lego"
model, since unnecessary elements can be removed or their stiffness could be a
numerical zero. However, this allows a first quantitative evaluation of the weight
of the new design (7.9 kg, 13.7% lighter than the original), its stiffness and

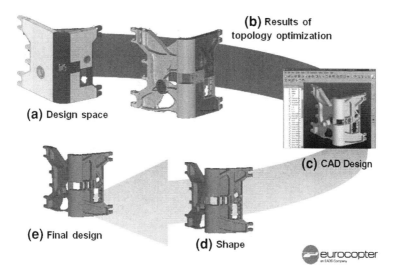

Fig. 6.11 Optimization-driven redesign

strength. On this basis, a new Computer-Aided Design—CAD model can be designed.

The new model is meshed and sized by using shape optimization. In order to define the node perturbation vectors, the FE model is parameterized with a morphing tool and a total of ten shape variables are defined. Basically they vary the thickness of the ribs and pockets.

Since shape variables can be specified in order to meet manufacturing requirements, no manufacturing constraints are necessary. The sized model has a mass of 7.46 kg.

This exercise was very successful. The two-step optimization-driven design process described above led to a superior support arm with a total weight-saving of 18.5%, while maintaining the original stiffness values and meeting strength and manufacturing requirements. These results were obtained in a two-to-three weeks timeframe, which is substantially less than that required by experienced engineers to design the original support arm.

6.3.2 A350 Fuselage Tail Section

In 2004, EADS-Military Aircraft competed for the acquisition of a work package for the new fuselage tail section 19 for the Airbus A350. The proposal was a new aluminum–lithium concept derived from the A330-200 with a reduction in weight. As part of the acquisition process, a concept study was carried out using topology optimization to suggest optimal inner structure within the tail cone design space.

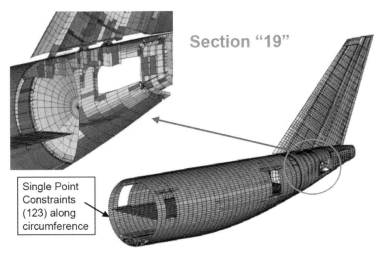

Fig. 6.12 A330-200 Global FE model used as baseline

Based on this, one innovative structural concept was derived and sized with a preliminary free-sizing optimization.[5]

The main functions of the fuselage tail section 19 of a passenger aircraft are to connect the horizontal and vertical tail planes to the pressurized cabin structure, to provide aerodynamic shape and to sustain the auxiliary power unit in the tail cone (fuselage section 19.1). High loads are transferred from the horizontal stabilizer that is hinged to the section 19 and trimmed by means of a screw in the central area. The vertical stabilizer is connected to the section 19 by means of six lugs. As a baseline for A350 section 19 concept development, a previous A330-200 was used. First, the rear fuselage of the A330-200 global FE model was constrained behind the wing, far enough from the area of interest in order to avoid perturbation to the load distribution due to constraints (Fig. 6.12).

The design space was limited by the outer skin and by the cut-outs for the horizontal stabilizer and the trim screw. The topology optimization problem was formulated as a minimization of the total weighted compliance for a given mass. More than 130 load cases of A330-200 were included in the evaluation of the weighted compliance. These included different cruise, maneuver, emergency and crash load cases. In order to evaluate trade-offs, different constraints for the available mass were used: the baseline A330-200 section 19 FE mass (M_0), M_0—10%, M_0—20%.

Furthermore, the thickness distribution of the outer skin was optimized. These sizing variables were used to update the stiffness of the outer skin following the change of internal loads during topology optimization of the inner structure.

---

[5] Engineers from EADS-Military Aircraft and Altair Engineering performed the topology optimization and reengineered the results, while the author performed a preliminary sizing. This study was presented in [27]. For this project the software package HyperWorks [28] was used.

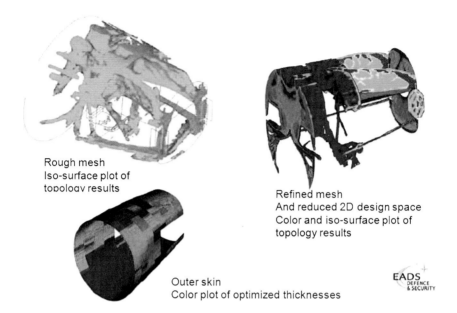

Rough mesh
Iso-surface plot of
topology results

Refined mesh
And reduced 2D design space
Color and iso-surface plot of
topology results

Outer skin
Color plot of optimized thicknesses

EADS
DEFENCE
& SECURITY

Fig. 6.13 Section 19 topology optimization

Figure 6.13 shows the results of the topology optimization of the internal structure (coarse 3-D mesh and finer 3-D/2-D mesh) as well as the optimized thickness distribution of the outer skin.

The optimal material distribution was carefully interpreted by experienced engineers and a new concept was derived. This is shown in Fig. 6.14 in comparison with the A330-200 frames. In comparison to the A330-200 design, this new concept shows new shear walls that connect the load introduction points of the horizontal stabilizer directly to those of the vertical stabilizer. Furthermore, the number of frames is reduced and efficiently placed in correspondence with the vertical stabilizer lugs.

In order to quantify the real benefits in terms of weight savings, the concept had to be sized. Because of scheduling pressure, no detailed sizing was possible. Free-sizing optimization was therefore used: the new CAD model was meshed with a fine shell mesh and built into the FE model of the A330-200. The optimization problem was defined as in the previous case: the weighted compliance was minimized with the same mass constraints (M_0, M_0—10%, M_0—20%). The idea was to obtain three sized concepts for stiffness in this way, while strength was evaluated by comparing the stress level in the outer skin and the internal structure with the A330-200 baseline. Stability was not considered directly in the optimization, but was qualitatively assessed.

The stress plots of the sized structure for the most critical load case are shown in Fig. 6.15. A greater efficiency of the new structures is clearly demonstrated by a lower stress level, even with a 10% reduction in mass. Furthermore, the stresses

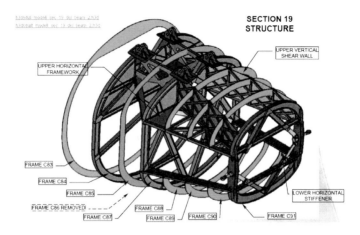

Fig. 6.14 Section 19, reengineered concept design (yellow) vs. baseline (blue)

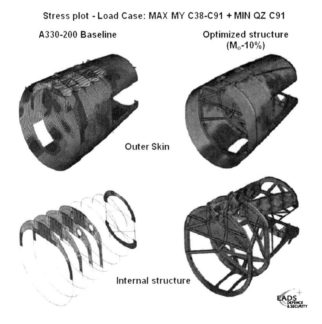

Fig. 6.15 Section 19, stress plot

are much more homogeneous than in the baseline, and peaks are reduced. This also results in expected improved fatigue life.

Clearly, stiffness was the driving design requirement. Figure 6.16 shows for three important load cases (LC) that the relative horizontal deflection at the vertical stabilizer tip was reduced, and stiffness therefore increased, relative to the

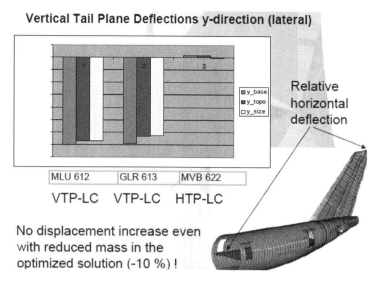

Fig. 6.16 Section 19, stiffness comparison

A330-200. Comparing the stress level and the stiffness of the different sized variants, EADS has estimated it can deliver a feasible design with possible weight, with savings between 15 and 20% with respect to the baseline of the A330-200. The duration of this conceptual study cannot be published, but it has shown to be compatible with the project scheduling.

6.4 Conclusions

Computer simulation and in particular FEA have been a vital part of the aircraft design process for a long time. It has been shown here how simulation has a double function: to support engineering in the early design phases and to validate the design in the detailed design phase. As state-of–the-art of simulation software evolves and computational resources increase, these functions are evolving, providing an important contribution to the improvement of the design process.

Regarding the engineering function, it has been seen that numerical structural optimization is establishing itself as a powerful tool to assist the design process. Analyzing two examples with different complexity, namely the application of numerical structural optimization to a simple bracket and to a fuselage section, it has been shown how numerical structural optimization helps to mitigate the design paradox (Figs. 6.2, 6.3). For example the application on the section 19 resulted in a fully-sized concept very early in the process, with a clear contribution to risk mitigation, helps in decision-making (e.g. benchmarking this concept versus a composite one) and shifting of knowledge (e.g. making available a high-fidelity weight estimation) upfront.

In addition, the more traditional validation function is evolving. The trends are showing on one side the necessity to standardize and harmonize methods and tools in the extended enterprise, and on the other, a constant pressure to reduce time and costs. Applying a highly-automated justification process, for example, is currently possible for a variety of similar components such as cable halters and brackets as the illustration provided shows. This application not only leads to a great reduction in the time required for the justification, but also provides a standardized process to be deployed in the extended enterprise with clear benefits in terms of quality.

References

1. Airbus evaluates heavier A380 (2003) Flight International Magazine
2. http://en.wikipedia.org/wiki/Airbus_A380. Accessed 1 February 2011
3. http://en.wikipedia.org/wiki/Airbus_A350. Accessed 1 February 2011
4. "Boeing Postpones 787 First Flight". http://boeing.mediaroom.com/index.php?s=43&item=720. Accessed 1 February 2011
5. "NASA Press Release 2008", http://www.nasa.gov/centers/dryden/news/X-Press/stories/2008/10_08_technology.html. Accessed 1 February 2011
6. Bruhn EF (1973) Analysis and design of flight vehicle structures. Jacobs Publishing Inc
7. HSB, Handbuch Struktur Berechnung, 12571–01, S.3, DASA-AIRBUS, 1995 http://www.fatec-engineering.com/downloads/HSB.pdf. Accessed 1 February 2011
8. Niu MCY (2005) Airframe stress analysis & sizing. Technical book company. Tabernash, CO, USA
9. Riviere A (2004) Gestion de configuration et des modifications lors du dévelopment de grands produits complexes en ingénierie concourante – Cas d'application aéronautique, PhD Thesis, Istitut National Polytechnique de Grenoble
10. Roskam J (2002) Airplane design: Airplane cost estimation: Design, development, manufacturing and operating (Airplane Design Part VIII), Darcorporation
11. VIVACE project, http://www.vivaceproject.com. Accessed 1 February 2011
12. Haftka RT, Gurdal Z (1992) Elements of structural optimization, 3rd edition. Springer
13. Schuhmacher G, Murra I, Wang L, Laxander A, O'Leary OJ, Herold M (2002) Multidisciplinary design optimization of a regional aircraft wing box. In: Proceedings of the 9th AIAA/ISSMO symposium on multidisciplinary analysis and optimization, Atlanta, Georgia
14. Schuhmacher G, Stettner M, Zotemantel R, O'Leary OJ, Wagner M (2004) Optimization assisted structural design of a new military transport aircraft. AIAA-2004-4641. In: Proceedings of the 10th AIAA/ISSMO symposium on multidisciplinary analysis and optimization conference, Albany, NY, USA
15. Töpfer G, Schäfer M, Schmidt J, Plönnigs J, Rudolph M (1999) Programmsystem Festigkeit 2000. DLR-Interner Bericht. 131-99/43. 63 S
16. JEC Compositecs, http://www.jeccomposites.com/composites-news/4882/software.html. Accessed 1 February 2011
17. Arnaudeau F, Mahé M, Deletombe E, Le Page F (2002) Crashworthiness of aircraft composites structures. In: Proceedings of IMECE2002 ASME international mechanical engineering congress & exposition New Orleans, Louisiana, 17–22 Nov 2002
18. Timoshenko SP, Gere JM (2009) Theory of elastic stability. Dover Publications, Mineola, NY, USA

19. Xiang Y (2002) Buckling of triangular plates with elastic edge constraints. Acta Mech 156(1–2):63–77
20. Advanced aerospace materials: past, present and future, Aviation and the Environment, 03/09, http://users.ox.ac.uk/~ smit0008/Publications_files/ORI-Aviation-Materials-2009.pdf. Accessed 1 February 2011
21. Altair Smashes Full-Vehicle Crash Simulation Time Barrier: CAD-to-Results in Less Than 24 Hours. http://www.altair.com/newsdetail.aspx?news_id=10398&news_country=en-US. Accessed 1 February 2011
22. Cervellera P (2007) Integration of structural optimization into the aircraft structural design process at subsystem and component level, PhD Thesis (tutor: Prof. U. Cugini), Università degli Studi di Padova
23. Zell D (2010) Optimization strategies applied to advanced ariane 5 upper stage concepts with specific focus to mass reduction of stringer stiffened cylinder shells. In: Proceedings of the European hyperworks technology conference 2010, Paris
24. Bruns A (2010) Sizing-Optimization of different flap-concepts. In: Proceedings of the european hyperworks technology conference 2010, Paris
25. Bautz B (2010) Optimization based concept development of innovative cfrp cargo-floor structures. In: Proceedings of the european hyperworks technology conference 2010, Paris
26. Cervellera P, Cabrele S, Machunze W (2009) Benefits of structural optimization in the aerospace design process: recent industrial applications on composites structures. In: Proceedings of 2009 NAFEMS World Congress, Crete, Greece
27. Schumacher G (2005) A350 Fuselage tail section 19 – Concept design by optimization methods, OptiStruct users' meeting, Stuttgart, Germany
28. Altair HyperWorks 2006: Users guide. Altair engineering Inc., http://www.altair.com. Accessed 1 February 2011

Chapter 7
Product Virtualization: An Effective Method for the Evaluation of Concept Design of New Products

Monica Bordegoni

Abstract This chapter discusses about the methods, tools and issues related to the practise of Virtual Prototyping used in the product development process. Virtual Prototyping is becoming more and more a diffused practise in various industrial sectors. Virtual Prototypes can be effectively used to validate the design solutions, already in the early phase of product design, when the engineering of the product is in the early phase or even not started. This practice can be used for checking the correspondence of the concept design with the users needs', and also for checking the users' acceptance of the new product through tests performed directly with end users'. This chapter is focused on Virtual Prototyping applied to consumer products, i.e., products that are characterized by aesthetics aspects, and by the fact that the users' interact with them. The chapter includes two examples of prototypes that have been developed with different objectives, and using different technologies (one in based on Virtual Reality and the other one on Mixed Reality technologies), and discusses the issues and the benefits.

7.1 Introduction

In the evolving scenario of the global market, and in a period of global crisis for both industrialized and non-industrialized countries, companies need to remain competitive, to consolidate their position on the market and, moreover, to strengthen one's leadership. While in the past, companies developed products with the aim of properly responding to customers' needs in order to improve their probability of success on the market, at the present time this is not enough. Today,

M. Bordegoni (✉)
Dipartimento di Meccanica, Politecnico di Milano, Via La Masa 1, 20156 Milano, Italy
e-mail: monica.bordegoni@polimi.it

M. Bordegoni and C. Rizzi (eds.), *Innovation in Product Design*,
DOI: 10.1007/978-0-85729-775-4_7, © Springer-Verlag London Limited 2011

the fulfillment of the customers needs is one of the basic strategic approaches in the product development process, while the full satisfaction of customers is becoming more and more a competitive feature of new companies, and a key element of business strategy.

The analysis and mining of what customers really want, capturing the so-called Voice of the Customer (VOC), is a strategic and fundamental source of information regarding product features. Traditionally, VOC surveys are usually carried out by marketing departments of the companies, without the involvement of the engineers and designers.

Then, in the first step of the product development process the customers' needs and wishes collected by means of various techniques are transformed into product requirements that can be acted upon by designers, who develop the concept of the product by transforming the requirements into specifications of the product. This phase is critical since the creativity and sensibility of designers intervene in the interpretation of users' needs, and in their representation into a *concept design*. Checking if the design well implements the users' needs is therefore crucial and is part of the subsequent activities. The first evaluator of the design once completed is the designer himself. But usually the more valuable suggestions come from the future users of the product. The designer can test the aesthetic, functional, technical and performance aspects of the products, but end users, who are the future potential customers of the product, can validate the acceptance of the concept design, and assess those aspects related to ergonomics and usability.

The evaluation of the concept of a new product is today made late in the traditional design process, after the product has been designed in detail (after the embodiment and detailed design) [1–3], and is used to identify the most satisfactory concept properties which at best meet the requirements previously identified. Then, the product concept is eventually modified in line with the input collected in the evaluation phase. This approach implies that the possible changes that can be made to the selected concept defined in this phase can be very limited because the most important design decisions about the product architecture and components have already been taken. On the contrary, it has long been established that conducting product evaluations early in the development process reduces the cost associated with making changes, and contribute in making better products [4].

The method of presentation of the concept is important, since it has to be well-understood by common users. Several methods can be used, as drawing, cardboards, mock-ups, physical prototypes or virtual prototypes. The method of presentation should be selected on the basis of the test purpose. Also the fidelity of the presentation, defined as the degree of closeness to the intended product, depends on the tests purpose, and on the moment of the design process when it is developed [5]. While some aspects, as aesthetic aspects, and perceived visual quality, can be today effectively assessed using virtual prototypes, all those aspects related to the use of the products (ergonomics, usability, physical interaction and response) still require physical prototypes.

Anyway, recently the use of Virtual Prototypes (VPs) in the product development process is becoming more and more a diffused practice [6]. This is due to the

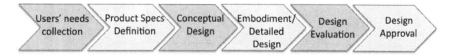

Fig. 7.1 Main phases of the product design process, including the phase related to the evaluation of the product design

fact that once created, the VPs are easy-to modify, share, represent in different ways and enriched with additional information. All these have often a reduced cost if compared with the same tasks performed on physical prototypes.

The initial use of VPs was constrained by the available technology, which was mainly limited to visualization systems [7]. Till few years ago the evaluation of a new product was performed in a virtual testing environment mainly based on high-performing visualization capabilities. These environments are appropriate for the evaluation of the aesthetic aspects of products, as variants of colors, textures and materials, but not for the evaluation of other features, especially those related to ergonomics and usability. The possibility of physically touching and manipulating, and also operating and interacting with a product, is an important issue in the assessment of several products, mostly of those that are interactive in their nature. Obviously, the evaluation of physical interaction through "non-physical" media is definitely poor and does not generally provide useful results.

In support of this, Virtual Reality and Augmented Reality technologies are evolving and improving very quickly, as well as are diminishing their costs [8]. For example, haptic technologies allow us to physically interact with virtual products, sonification techniques allow us to hear sounds emitted by products during operation in a realistic way. This trend encourages the development of VPs of industrial products, which are increasingly performing, and similar to the corresponding real products. Therefore, more and more VPs allow us to evaluate various aspects of the design of new products with a good level of fidelity. We can check the aesthetics of a product, also placed in a use context, and we can also evaluate its ergonomics and usability aspects.

The chapter is focused on the use of VPs to assess the concept design of new products. Figure 7.1 shows the traditional main phases of the product design process, which includes the evaluation of product design performed after the embodiment and detailed design.

In this chapter we show how VPs can be effectively used to validate the design solutions, in the early phase of product design, when the engineering of the product is also in the early phase or even not started (Fig. 7.2). This practise can be used for checking the correspondence of the concept design with the users' needs, and also for checking the users' acceptance of the new product through tests performed directly with end users. Specifically, the chapter focuses on Virtual Prototyping applied to consumer products, i.e. products that are characterized by aesthetics aspects, and by the fact that the users interact with them.

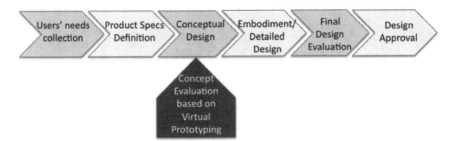

Fig. 7.2 Main phases of the product design process, including the evaluation of the concept design based on the use of virtual prototypes

7.2 From Users' Needs Collection to Concept Design

The main objective of the survey of users' needs is to capture all the features of the product that could be relevant for the users, in order to obtain information to create functional and technological requirements, design specifications, psychological and emotional aspects and so on.

Users' needs, also named the Voice of the Customer (VOC) can be captured in several ways, as customer interviews, customer surveys, focus groups, ethnographic studies, etc. [9]. Specifically, the techniques most commonly used are mail-out, in-person interviews and telephone surveys. Traditionally, these kinds of VOC survey are usually carried out by marketing departments of the companies without the involvement of the engineers and designers. Nowadays, this approach has showed its limits related to the distortions of the translation of the voice of the customers into the product requirements made by marketing people. Therefore, at the present moment the trend is moving toward the involvement of industrial designers and engineers in researching customer needs, through meeting with customers, observing customers using products and participating in focus groups.

Despite these attempts to improve the VOC survey methods, some issues still remain uncovered: these are related to the difficulty to collect data about "unstated" or "unspoken" needs and wishes, which could represent the most important input for defining product requirements.

Once users' needs are collected, then they have to be analyzed and organized in a general overview of preferences. Then, the needs and wishes collected are transformed into *product requirements*, which are more technical expressions of user needs: these consist of a list of requirements that represent the needs and the desired outcomes of the users who interact and use the product. The Quality Function Deployment (QFD) is one of the methodologies, which is used to support this process of transformation [10].

In the subsequent step of the product development process the product requirements are transformed into design parameters, which are the most relevant elements of the product that aim at satisfying the requirements firstly defined. This transformation depends on one side on the completeness and accuracy of the data captured at the beginning with the users needs collection, and on the other side requires a good understanding of different disciplines as technology, science, engineering, design and so on, that have to be applied also in a correctly structured product development process.

One approach to users' needs collection and transformation into product parameters is proposed by the Kansei engineering. Kansei Engineering was created to address the emotional side of product development [11]. It is a consumer-oriented emotional design and branding methodology that attaches importance to customers' wants for emotional, sensory and lifestyle-enhancing solutions with product development, creating synergy for emotional branding. The main issues of this technique concern the methods used in order to grasp the user's feeling (kansei) about the product, the identification of the design characteristics of the product from the users' kansei, the various techniques adopted to build Kansei Engineering systems and the methods used to adjust the product design to the current societal change in people's preferences [12].

At the end of this conceptual phase, users' needs and wishes are transformed into functional, psychological, technological, emotional and communicative parameters of the product. On this basis, the proper product design phase starts, including embodiment and detailed design. Product design can be defined as the process by which an idea is transformed into a tangible outcome taking into account and satisfying objectives, constraints and needs.

In this chapter we are interested in investigating conceptual design, which is the phase when concepts are generated with a view to fulfilling the objective. This phase includes comparing alternatives to select the best possible solutions. Therefore, it is mainly about exploring and experimenting.

One of the big concerns for many companies is how to generate more and better ideas, and how to become more creative in the proposed product. Workload, pressure and time constraints available for the development of new ideas may be obstacles to creativity and innovation, as well as the impossibility of evaluating and experimenting with various solutions and alternatives in search of the best may limit the potential success of the future products.

7.3 Design Evaluation and Prototyping

There is great uncertainty as to whether a new design will actually do what is desired or will be accepted by users. In order to be effective, a design should satisfy users' expectations from at least three points of view: aesthetics, functionality and ergonomics [1]. It is not so uncommon that users' tests reveal design errors or misinterpretations of the initial provided requirements. Generally, an accurate check of user acceptance enables a better design of products, and eventually better and more accepted products. During the tests, some selected representations of the concept are submitted to groups of target users for the evaluation.

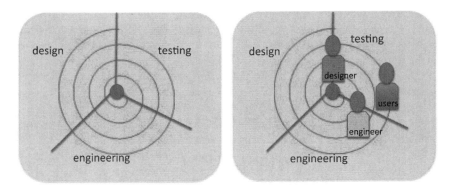

Fig. 7.3 Design evaluation phases and persons involved. For an effective validation of the product concept design and use of the test results for improving and optimizing the design, the design team should be directly involved in the testing phase

We distinguish three major phases—which are design, engineering and testing, in the design process, which are iterative in their nature, since the outcome of tests, if revealing a non-fully satisfactory design, restarts the process. Multiple iterations of tests are used to progressively refine the design. Hereafter, we represent the process as a spiral, finally converging to the optimal description of the product (Fig. 7.3). In this process, the designers are involved in the conceptual design phase, the engineers in the product engineering and detailing, and the users in the testing phase. In order to be effective the design team—including designers and engineers—should also be directly involved in the testing phase.

7.4 Prototyping of Products

In general, it is easier to understand a product concept if there is a realistic representation and a physical model to see and use. Therefore, a prototype is often used as part of the product design process to allow the design team to explore design alternatives, to test theories and confirm performance prior to starting the production of a new product.

Prototyping is a common practice for evaluating globally and in an integrated way the design of the product under development, and for correcting misunderstanding and errors at the early conceptual design phase. Although prototyping is often considered too expensive, correcting ambiguities and misunderstandings at the specification stage is significantly cheaper than correcting a product after it has gone into production.

Different types of prototypes can be used according to the test purpose. The implementation with which a prototype represents a product can also vary,

Table 7.1 Aspects of interactive products that can be evaluated by the users through the various senses

Aspects to evaluate	Human senses		
	Sight	Touch	Hearing
Aesthetics			
Shape	X	X	
Color	X		
Material	X	X	(X)
Ergonomics/Usability			
Affordance	X		
Accessibility	(X)	X	
Visibility	X		
Sensorial feedback	X	X	X
Easiness of use		X	
Perceived comfort	X	X	X

according to the needs and available technologies. The level of the fidelity is related to the needs and objectives and even to the maturity of the concept development. Of course the prototype is not the final product, therefore there are some limits to consider during its use in the tests. All these issues are discussed in the following sections.

7.4.1 Product Aspects to Evaluate

Typically in products there are some aspects that we are interested to evaluate with the involvement of users. These aspects are mainly related to the aesthetics features of the products, as the shape, color and material, and to the ergonomics and usability aspects, as the affordance, the accessibility, the visibility, the easiness of use, the sensorial feedback and others (see Table 7.1).

These aspects are generally evaluated by the users through their senses: typically the sense of sight, of touch and of hearing. For example, the color can be evaluated visually, the material can be evaluated both visually and through the sense of touch. The sensorial feedback involves all senses, while visibility only involves the sense of sight.

7.4.2 Types of Prototypes and Test Purposes

There are several types of prototypes, which can be related to their functions and purposes in the tests. Prototypes can include all the properties of the product or only few of them. We can organize tests for evaluating one property at a time, which is detached from the others, or we can adopt a more "holistic" approach, where all the properties of a product can be tested at the same time.

We have identified the following kinds of prototypes.

VISUAL prototypes. They show the intended design aesthetics and simulate the appearance, color and surface texture of the intended product. They do not embody the functions of the final product.

SHAPE prototypes. They show size and form of a product. They are used to assess visual aspects and ergonomics factors of the product final form. Actually, they do not simulate the exact visual appearance (color, finish or texture) and product functions. Often, they are hand-made models, Rapid Prototyping models or machined models.

FUNCTIONAL prototypes. They are built with the aim to simulate the functional and operational principles of products.

FULL PHYSICAL prototypes. They simulate the final design, aesthetics, materials, functionality of the intended design. Usually, they are fully working scaled prototypes, or even full-scale prototypes.

7.4.3 Prototypes Implementations

Prototyping has various implementations [1]. In the industrial product sector, prototyping has been traditionally intended as physical prototyping.

Physical prototyping plans the development of a partially or fully-working product, which is in general a too expensive activity to carry out in the early stage of design; it is more suitable to be performed at the end, when most of the design problems have been identified, addressed and potentially solved. But this practice, which offers the advantage of physically interacting and validating the design proposals, is possible only if a full 3-D digital model is available, and sometimes with limitations in terms of size and materials (for example, in the automotive sector).

Rapid prototyping is another interesting opportunity [13]. It is fast and cheaper, but less comprehensive for what concerns the product features that are necessary for properly evaluating the effectiveness of the design. Nevertheless, new technologies for Rapid Prototyping are becoming widely used and distributed. Very recently Rapid Prototyping machines have been commercialized for the personal production of objects [14, 15].

Prototypes and rapid prototypes have some major problems: the first are expensive and the latter are not rich enough for a comprehensive validation of the various relevant aspects of the product. For these reasons, the use of *Virtual Prototyping* is a practise that is spreading in the industrial design and engineering fields.

Table 7.2 Comparison of characteristics of prototypes and final products

	PROTOTYPE	FINAL PRODUCT
FIDELITY	-Less details -Require refinement	- Full details
MATERIALS	- Simulation of intended final material	- High quality materials and finishing
PRODUCTION PROCESS	-Flexible/low cost manufacturing process	- Time consuming and costly process

Virtual prototyping aims at effectively supporting the validation of the initial concept of new products, while leaving the full validation of the complete product at the end of the design process, through a fully-working prototype (as shown in the product design process of Fig. 7.2). With this new practice, it is possible to perform design review before building hardware products. Consequently, the number of physical prototypes effectively built during the overall product development process is considerably reduced. And in addition, more variants and iterations are possible, as well as a direct comparison of the various variants.

7.4.4 Fidelity of Prototypes

Prototypes can be built with a variable level of fidelity, according to needs and purposes, and also according to the maturity of the product concept development. Fidelity is referred to the degree of closeness to the intended product [5]. The concept of prototype fidelity is defined by the level of detail used in making the prototype, and thus it indicates how closely the prototype resembles the "real object". For example, low-fidelity prototyping can be achieved with simple tools such as paper and colored pencils with which quick and rough sketches of the product can be drawn.

On one side the goal of users' tests is to make users' responses to the presentation of the concept the same as to the real object. Conversely, making very realistic prototypes can be quite expensive and time consuming. Therefore, the selection of the type of prototype depends on how much the prototype should resemble the final version of the design. Of course, the level of fidelity can affect the evaluation of the design as well as the emotional involvement and responses of the users. Low-fidelity prototyping can represent a work in progress, having limited details. So it can be a modality for representing our concept that is cheap and simple, as well as powerful.

7.4.5 Prototypes and Products

Obviously, there is a difference between a prototype and the final product. These differences, which are summarized in the following table (Table 7.2), are mainly related to three aspects: fidelity, materials and production processes.

The fidelity of a prototype usually includes less detail and requires refinements with respect to a product that is more complete and includes full details. For what concerns the materials, usually the prototype is manufactured using a simulation of the intended final material, instead the final product is produced by means of traditional manufacturing processes and is made of better quality materials and finishing. Finally, the production process is different: prototypes are produced paying attention to the production costs, while final products are manufactured using time-consuming and costly-production process, where the attention is paid to the quality.

7.5 Virtual Prototyping

Virtual Prototyping is an anticipation of a product that does not exist in reality yet. We call it "product-to-be". Designers, engineers and also end users can use a virtual prototype for evaluating a product-to-be aesthetic quality, its functional aspects and also ergonomics and usability aspects. Several successful stories about using virtual prototyping for design assessment are reported in the literature [16–19].

Virtual Prototyping is related to the construction and testing of virtual products, which is based on the use of Virtual Reality (VR) technologies. Most advanced VR technologies include highly-realistic visualization techniques and technologies, and multi-modal interaction modalities that are based on vision, haptic and auditory modalities. Therefore, it is clear that today the implementation of a Virtual Prototype requires a multi-disciplinary perspective including technical disciplines (Computer Science, Computer Graphics, Computer Vision, Mechanical Engineering, Electronics), Human-related disciplines (cognitive psychology, ergonomics) and Industrial Design.

Till few years ago the virtual testing of a product-to-be was performed in a virtual environment mainly based on high-performing visualization capabilities. Basically, VPs consisted of a realistic visualization of the product. Today, advanced stereoscopic visualization technologies allow for a very realistic, often high-fidelity and real-time, visual perception of the virtual model of the product [20]. This practise is appropriate for the evaluation of aesthetic aspects of a product, as variants of colors, textures and materials in different lighting conditions, but not for the evaluation of other features, especially those related to the use of the products.

The possibility of physically touching and manipulating, and also operating and interacting with a product model, is an important issue in the assessment of several products, especially those products that are interactive in their nature. In fact, the evaluation of the physical interaction with virtual components, that are actually "non-physical", is definitely poor and does not provide useful results in terms of product evaluation.

7.5.1 Enabling Technology

The technology enabling the development of VPs is in the domain of VR. VR technologies allow us to represent worlds that exist in reality, with various degrees of realism, and also that do not exist in reality. VR technologies have expanded with the introduction of Augmented and Mixed Reality technologies, where the world perceived by the user is not completely virtual, but where virtual and real objects are integrated and mixed together. This section shortly describes the most common technologies that are at the basis of the development of VPs.

7.5.1.1 Virtual Reality Technology

Virtual Reality (VR) is referred to computer-simulated environments that can simulate places and objects in the real world, as well as in imaginary worlds [21, 22]. The simulated environment can be similar to the real world (i.e., in simulations for pilot training) or can differ significantly from reality (i.e., VR games). Most VR environments are primarily based on visual experiences, displayed either on a computer screen or through stereoscopic displays, immersive HMD and CAVE. But today some VR applications also include additional sensory information: force-feedback and tactile information through haptic devices, as well as sound rendered through speakers or headphones.

VR applications are interactive. The users can interact with virtual places and objects through multi-modal devices: haptic devices, wired glove, Wii mote, tracking systems, etc. Particularly effective is multi-modal interaction, which is obtained through a combination of various user input modes, such as speech, touch, manual gestures, gaze and head and body movements.

In recent years, the sense of touch has been integrated into computer-based applications through the **haptic technology**. Haptic devices are divided in two main categories: tactile displays and force-feedback devices [23]. The tactile displays are those haptic devices that exploit the modalities of the skin sensors. These systems can be classified into the following major categories based on: static pressure or vibration (mechanical energy), electric field and temperature difference (or thermal flow). Force-feedback devices generally return forces and torques to users' hands. There are several force-feedback devices available on the market; some of the most popular ones are the Sensable Phantom [24], the MOOG-HapticMaster [25] and those distributed by Haption [26], which are basically general purpose and point-based devices, i.e. they allow us to simulate the effect of a single contact point between the users' hand and a generic virtual object. Some other devices, which are currently just laboratories' prototypes, have been developed returning a full-hand and linear contact-type or surface contact-type with the virtual object. Two research projects have recently investigated the creation, modification and exploration of digital aesthetic surfaces through the use of novel haptic interfaces: the *T'nD project* [27] and the *SATIN project* [28, 29]. Besides, there are ad hoc haptic devices, such as those developed for the medical applications [30] or programable devices like the haptic knob as described in [31].

The role of sound in product design has been neglected for a long time. In the realm of human factors, auditory warnings and alarms have been extensively studied since the nineties [32]. A community for **auditory display** (www.icad.org) was established in 1992, and concepts such as auditory icons, earcons and sonification became widely known. However, the design of sounds in interaction with products has become the subject of research efforts only in the last decade. In recent years, special attention has been devoted to sound as a means to provide affordances and feedback for continuous interaction with everyday objects [33, 34]. This has been made possible by the availability of cheap sensors, actuators and processors, such that the embedding of synthetic sounds within tangible objects becomes convenient. The development of efficient models for sound in everyday contexts has been spurred by the needs of interactive computer graphics, computer games and interaction design [35, 36]. Other approaches based on the modal description of objects [37], pre-computed with finite-element methods, can find their place in virtual prototyping to compare different object shapes, especially when coupled with software tools.

Virtual Prototyping based on pure visualization has demonstrated of being effective for the validation of aesthetic aspects of a product, but not demonstrated of having the same effectiveness for other kinds of evaluation, for example ergonomics and usability evaluations. In order to effectively substitute physical prototypes with virtual ones, and use them for these kinds of tests it is necessary to have a virtual model that has the same characteristics of, and behaves like the corresponding physical one for what concerns the aspects we are interested to simulate and validate. Several research works have demonstrated that a multi-modal approach in the interaction with VPs, i.e. the use of the combination of more sensory channels, is essential in order to achieve an effective and accurate validation of the characteristics of future products. For example, Bordegoni et al. [38] have demonstrated the effectiveness of using a multi-modal approach based on the combination of vision, sound and touch in creating a design review activity on continuously evolving shapes. In this application, industrial designers, who are manually skilled, have the possibility of interacting with the evolving shape of the product using their hands, and to continuously evaluate and modify the model. In [39] a multi-modal environment based on vision, sound and touch has been used to reproduce human interaction with the interface of a household appliance.

7.5.1.2 Augmented and Mixed Reality Technology

Interesting contributions to the field of Virtual Prototyping are coming from the Augmented and Mixed Reality domain [40]. Augmented Reality (AR) and in general Mixed Reality (MR) are two growing research topics within the VR research area, and address all those technologies that allow us to integrate virtual objects or information with the real environment [41]. AR has been defined by

Table 7.3 Technology for virtual reality applications and for augmented/mixed reality applications

Technology	VR applications			AR/MR applications		
	Visual	Haptic	Auditory	Visual	Haptic	Auditory
Power wall	X					
Immersive HMD	X					
CAVE	X					
Optical see-through HMD				X		
Video see-through HMD				X		
3DOF/6DOF general purpose force feedback device		X				
Ad hoc haptic device					X	
Tactile device		X			X	
Optical tracking system	X			X		
Pattern-based tracking system				X		
Headphones			X			
Loudspeakers						X

Milgram in 1994 as follows: "An AR system supplements the real world with virtual (computer—generated) objects that appear to coexist in the same space as the real world." [42]

The common implementation of AR/MR concerns the "augmented visualization", where virtual objects are inserted within images taken from the real world. This is basically accomplished by an analysis of the real environment that allows us to capture some geometrical information, and process it in order to align and integrate properly digital and real information. Real environments used for AR applications can be structured or unstructured, i.e. equipped with some fiducial markers and recognizable elements or not [43]. Regarding the technologies for visualizing merged real-virtual information it is possible to use common monitor displays, hand-held displays or optical or video see-through Head Mounted Displays (HMD). Up to now most of these devices are still academic prototypes [44] or are often developed for specific applications; but recently some ergonomics, low cost and performing goggles have been developed for the market [45].

AR/MR technologies have proved to be effective in the product design domain [46, 47]. Specifically, Mixed Prototyping is a flexible and powerful practice since it allows the interaction with both virtual and real prototype components, taking advantage of the possibility on one side of evaluating early in the design process components that do not exist already, and on the other side of better feeling components through physical interaction. This practice can be effectively used for the rapid design review of new products, as demonstrated in several research works: new design methods based on AR to support early design activities [48]; a method based on the combination of AR and RP for product development [49]; a

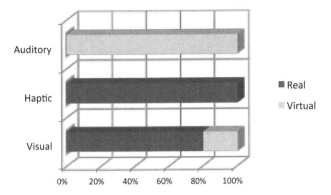

Fig. 7.4 Reference framework for virtual/augmented/mixed reality applications

MR application consisting of mock-up covered with designed virtual mock-ups for evaluating the designability and operability of products [50]; the use of VPs of information appliances for users' test in the early design stage, as well as some works carried out on the ergonomics assessment performed by using human models [51]; MR application for the evaluation of the concept of an information appliance, a video storage device, developed within a real design process [52].

Table 7.3 provides an overview of the typical technology used for the implementation of VR and of AR/MR applications, referred to the kind of interaction modalities—Visual, Haptic and Auditory–that the applications offer.

7.5.2 Reference Framework

It is proposed hereafter a reference framework consisting of a chart where we show which are the components of an application based on VR or AR/MR technologies. On the horizontal axis we represent the kind of components present in the application: there can be real components only, virtual components only or a mix of the two kinds. On the vertical axis we represent the kind of interaction modality, which can be visual, haptic or auditory. The chart reports the percentage of the component of a specific type that is present in the application and that is used for implementing a kind of interaction modality. The example shown in Fig. 7.4 refers to an application consisting of virtual sounds rendered through the auditory modality, a mix of real and virtual objects rendered through the visual modality. Only the real objects can be perceived through the sense of touch.

A MR application consists of a mix of real and virtual components: we can have a mix of virtual and real visualization of objects, we can touch and interact with virtual and real objects, as well as we can mix real and synthetic sounds.

7.5.3 Characteristics of a Virtual Prototype

The goodness and effectiveness of a virtual prototype can be evaluated in respect to several characteristics. In the following we list the main ones, which can be evaluated with respect to the three types of perceptual modalities: visual, haptic and auditory.

FIDELITY. This characteristic defines the level of fidelity of the virtual prototype in general, or of the various aspects of the product such as its visual representation, its haptic behavior, or its auditory features.

COMPLETENESS. This characteristic identifies which is the degree of implementation of the product design that the prototype covers. In fact, typically a prototype implements some parts, but not all, of the product design, also in relation to the kind of evaluation and test that we intend to carry out.

FLEXIBILITY. This characteristic identifies how easy and quick is to implement changes in the virtual prototype features. For example, it indicates if it is possible to change colors and textures of the interior of a car, or if is possible to change the haptic behavior of a door handle, quickly and easily as soon as a user asks for it.

COMPLEXITY of REALIZATION. This characteristic defines the complexity of developing the virtual prototype. It depends on the VR technologies that are used, on the ad hoc components that need to be specifically developed for the realization of the prototype, on the software code that has to be specifically implemented, etc.

7.6 Case Study: Virtualization of Interactive Products

This section describes two case studies consisting of two different kinds of VPs of a product equipped with interaction elements, which we have selected for our experiments. The product consists of a washing machine. Two replica of the washing machine have been developed by simulating the look and feel of some of the interaction components of the domestric appliance (for example, the knob for the wash programing) and by using an approach based on VR technologies in one prototype, and MR technologies for the implementation of the other one. The two virtual washing machines are discussed in terms of the characteristics of the VPs, and of the aspects of the product that we are interested to evaluate.

7.6.1 Product Virtualization with Virtual Reality Technology

The first case study consists in the development of the prototype of a washing machine based on the use of VR technologies. The aim of the prototype is to test with users the aesthetic features of the washing machine, and the ergonomics of its interaction components (door, buttons, etc.).

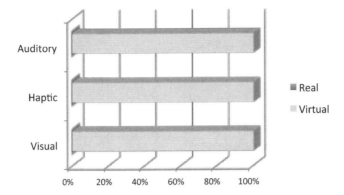

Fig. 7.5 Collocation of the virtual prototype in the reference framework

Fig. 7.6 Setup of the virtual washing machine based on VR technologies

Figure 7.5 shows the collocation of the prototype in the Reference Framework presented in Sect. 7.4.2. The prototype provides a highly-realistic visualization of the appearance of the washing machine, and of the haptic and auditory interaction with the elements of the washing machine and of its control panel. the door, the drawer, the knob and the buttons.

The development of the application is described in detail in [53], and is based on the following VR technologies:

- Haption Virtuose, with 6DOF to render forces and torques (www.haption.com);
- Cyviz, a retro-projected wall display for the stereoscopic rendering of the scale model of the appliance (www.cyviz.com);
- Three AR-Tracking cameras for the detection of the user's point of view (www.ar-tracking.de);
- A headset equipment for sound rendering;
- The 3-DVIA VirTools, which is a development environment (www.3dvia.com) that have been selected because well integrated with the 3-DVIA CAD tools that have been used for implementing the digital models.

Figure. 7.6 illustrates the setup used for the realization of the Virtual Prototype.

This virtual prototype consists of the stereoscopic visualization of a highly realistic representation of the washing machine and of its components. The user's head is tracked, so that the user point of view can be changed according to his movement within the tracking space. The haptic interaction has been implemented by adding force-feedback to the interaction elements of the digital washing machine, which are manipulated by the user through the use of the Haption device. The magnitude of the forces has been roughly computed from simplified mechanical laws derived from the CAD model, and then empirically adapted to match the real ones. In addition, we have added sound to the interaction components. A unique sound for each colliding part has been recorded (for example, the door). This has subsequently been manipulated with a sound tool in order to reproduce different effects. For example, some reverbs and a compression filter have been used in order to create a modification of the basic sound that is used for the realistic simulation effect of higher forces applied to the interaction elements.

The user wears a pair of stereo glasses and the audio headset; he stands in front of the wall display and handles the haptic device. He can look at the washing machine from different points of view, thanks to the tracking system connected with the stereo glasses. He uses the haptic device for interacting with the haptic components of the washing machine: he can turn the knob, push the buttons, open/close the drawer, etc. In case he does not like the haptic response provided by one of the components, he can ask to change it. For example, the reaction to the button pressing action can be set to be stronger.

7.6.1.1 Discussion

We can evaluate the four main characteristics of the virtual prototype, as presented in Sect. 7.4.3, with respect to the three types of interaction modalities. The evaluation is qualitative and is made on the basis of three values: low, medium and high (Table 7.4).

The *level of fidelity* of the visual representation is high, since the washing machine is seen in stereoscopy and from different points of view. We have performed a comparative study of a physical washing machine with the corresponding

Table 7.4 Characteristics of the virtual prototype

| | Interaction modality | | |
Characteristics of VP	Vision	Haptic	Auditory
Level of Fidelity	high	medium	high
Completeness	high	high	high
Flexibility in modification	high	high	high
Complexity of development	low	low	medium

virtual model, which is described in detail in [54]. The study has shown that the intrusiveness and the non-compliance of the kind of haptic device (the Haption device) with respect to the shape of the real interaction components does not allow us to perform an optimal evaluation and a fine tuning of the haptic parameters. And even if the use of this device gives us great flexibility so that we can simulate the interaction with any interaction element, including buttons, the most similar ones are those where the shape of the generic handle does not distract the users from the task they are completing. So, while the haptic response is very similar to the real one, the fidelity of the haptic modality for the use of the interaction components is not very high, since these components are operated through the end-effector of the haptic device (handle-shaped), and this prevents from a natural interaction. For what concerns the auditory feedback, it has been obtained through a manipulation of a basic recorded sound, and the resulting fidelity is quite good.

The *completeness*, i.e. the degree of implementation of the product design that the prototype covers is good in relation to the kind of evaluation and test that we intend to carry out. Specifically, we can evaluate visually the aesthetic properties of the washing machine. We can perceive the sensorial feedback through the all three senses during the manipulation of the interaction elements, and we can check the accessibility and the ease of use through the sense of touch. At the moment the prototype does not allow to physically touch the all washing machine in order to evaluate its shape. Actually, it would be possible to implement the haptic model of the washing machine body and allow users to feel it through the sense of touch. Of course, the kind of haptic contact would be anyway limited, since it is point-based due to the type of haptic device used.

The development of the virtual washing machine has been performed using commercial hardware and software components. The hardware components are general purpose technologies, performing but not very cheap, which have been programed for this specific application. The virtual washing machine has been implemented using the VirTools development environment that is based on a building-block paradigm. This is a very easy programing paradigm that allows us to easily configure the application parameters. Therefore, the prototype is very *flexible* in case we want to change on the fly the parameters that are related to vision, haptic or auditory interaction modalities. Besides, the application allows us to quickly configure the prototype according to the specific test that we intend to perform.

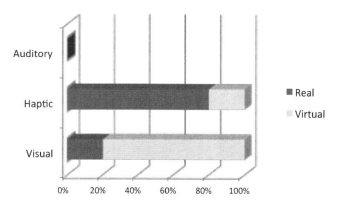

Fig. 7.7 Collocation of the MR prototype in the reference framework

The general *complexity* of the application development is low. In fact, VirTools supports the possibility of integrating various VR technologies (3-D visualization, tracking, interaction), and includes a library oriented to the physical simulation of effects. During the implementation, the main problem we faced was related to the fact that VirTools supports the integration of haptic devices, but has not been designed specifically for haptic interaction. So, it does not offer all the functionalities and the access to low-level control of the haptic devices that are usually available in other haptic libraries, such as CHAI3D or H3D. Then, the haptic effects that we have reproduced are obtained as the combination of a limited number of low-level force controls.

We have performed some tests with users with the aim of using the Virtual Prototype to set the preferred effects, so as to get the specification of an ideal product that users really like. During the experimental trials we have noticed that the intrusiveness of the interaction devices and the sense of presence influence the quality of the perceived information. In particular, we have seen how for small components, such as buttons, the shape of the end-effector has a negative influence on the quality of the interaction. While the quality of the physical reaction that we have simulated is appreciable, the grasping of a generic handle that pretends to simulate the shape of the interactive component is unsuccessful. Therefore, we come to the conclusion that it is important for performing effective usability and ergonomics tests to equip the haptic device with an end-effector that has exactly the shape of the real one. This can be solved easily by rapid prototyping the exact shape of the component (or a set of variants of the component), and mount it on the haptic device.

7.6.2 Product Virtualization with Mixed Reality Technology

The second case study consists in the development of the prototype of a washing machine based on the use of MR technologies. The aim of the prototype is to test

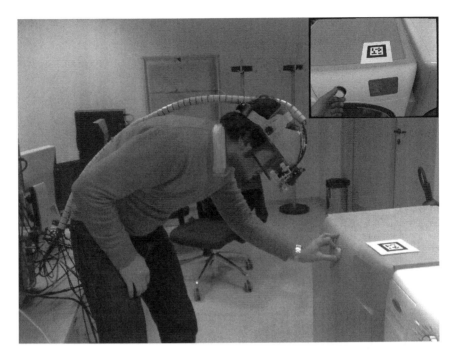

Fig. 7.8 Virtual washing machine based on MR technologies

with users the aesthetic features of the washing machine, and the ergonomics of the control knob.

Figure 7.7 shows the collocation of the prototype in the Reference Framework. The prototype consists of a highly realistic visualization of the appearance of the washing machine, perceived through a video-see-through HMD, and of the haptic interaction with the knob implemented through a specifically developed programable haptic device. The auditory modality is not implemented.

The development of the application is based on the following MR technologies:

- A video-see-through HMD (VST-HMD), through which the user can see the AR environment in stereoscopy [55];
- Programable haptic knob [31] specifically customized for this application;
- Marker-based tracking library ARToolKit [56];
- RTT DeltaGen software tool for real-time rendering (www.rtt.ag).

Figure 7.8 illustrates the setup used for the development of the AR prototype.

This application consists of a real-scale physical maquette of the washing machine, consisting of a box made of plywood, and a dashboard made with a Rapid Prototyping production technology, where the haptic knob is placed. A printed pattern is placed on top of the washing machine, so as to align the virtual washing machine with the physical box. The physical washing machine is

Table 7.5 Characteristics of the mixed prototype

Characteristics of VP	Interaction modality		
	Vision	Haptic	Auditory
Level of Fidelity	high	high	N/A
Completeness	high	high	N/A
Flexibility in modification	medium	high	N/A
Complexity of development	medium	medium	N/A

painted green in order to get an appropriate visualization of the user's hand interacting with the physical maquette, which is implemented through a chroma key technique [57]. The visual representation of the washing machine is implemented using the RTT tool, which provides a highly realistic real-time rendering, integrating the menu to program the wash, implemented using the Adobe Flash library (www.adobe.com). The knob is a programable device, in the sense that some of its functions can be changed and programed on the fly. For example, we can change the number of clicks, their amplitude, the maximum torque, etc.

The user wears the VST-HMD and looks at the physical washing machine: through the HMD he can see a realistic representation of the domestic appliance super-imposed onto the physical box. Then, he can operate the physical knob to change the washing program. He can also ask the application developer to change the haptic response of the knob, which can be done in real time.

7.6.2.1 Discussion

As in the previous study case, we evaluate the four main characteristics of the virtual prototype, as presented in Sect. 7.4.3, with respect to the three types of interaction modalities (Table 7.5).

The *level of fidelity* of the visual representation is high, in the sense that the washing machine is seen in stereoscopy and in the real context through the HMD. Conversely, some details of the control panel cannot be seen well, due to the low resolution of the display system. Also the fidelity of the haptic modality is high. In fact, the programable haptic device is an exact representation of the real control device, for what concerns both its shape and behavior.

The *completeness* is good in relation to the kind of evaluation and test that it was intended to carry out. In fact, this prototype was aimed to be used to test the aesthetic features of the appliance and the interaction with the knob device.

The rendering of the washing machine has been developed using the RTT tool, and the GUI of the control panel using the Adobe Flash tool. Due to the complexity of the tools, the modification of the virtual models is not straightforward. Therefore, the flexibility in changing the aesthetics and the GUI is not very high.

Instead, the haptic knob is a very flexible device, which is accompanied by an easy-to-use web application that allows us to change the force-feedback parameters [31].

The development of the MR washing machine has been performed using ad hoc hardware and commercial software components. The ad hoc hardware component has been previously developed and adapted specifically for this application. We have built the covering of the mechanism of the knob by using a Rapid Prototyping production technique. The behavior of the knob has been programed using a web interface. Of course, the development complexity increases in case ad hoc haptic components need to be developed and integrated into the application.

We have performed some tests with users with the aim of using the MR prototype for checking the aesthetic and the ergonomic features of the knob. The tests have revealed the fact that the users liked the approach, also because it was new, attractive and engaging. The realism of the simulation, both visual and haptic, has been rated highly by the testers. The alignment of the physical and virtual prototype is good, and the quality and resolution of the graphical information rendered is acceptable. The major issue that has been raised by the testers is about the discomfort when wearing the HMD. This technology is perceived as intrusive, and also limited for what concerns the possibility of moving and also for what concerns the field of view. Research on HMD devices is progressing fast, and we expect to be able to use very light, performing and cheap stereo goggles in the near future [45].

7.7 Discussion and Conclusions

The chapter has presented some issues related to the virtualization of products, a practice that is becoming commonly used for testing products design early in the concept phase. Virtual Prototyping of any products is effective if we clearly define the aim of the prototype, and what we want to test with users. Then, it is necessary to develop the prototype, which can require the use of VR or AR/MR technologies commercially available, or the implementation of new ones.

The chapter has presented a set of characteristics of VPs. First we consider the level of fidelity, which can vary according to the purpose of the prototype, as well as with the available technologies. Of course, if the level of fidelity is low with respect to the aim of the tests, this will limit the effectiveness of the tests. The completeness of the prototype is also important, again in respect to the aim. The flexibility of the prototype is a crucial parameter, since it is related to the possibility of easily configuring the prototype characteristics with respect to the kinds of tests and of the users. Finally, the implementation can be complex and can take time. But generally, the benefits obtained by the flexibility offered by using a Virtual Prototype for the users' tests are greater than the time and costs related to its implementation.

The chapter reports two examples of prototypes that have been developed with different objectives, and using different technologies (one uses VR and the other one uses MR technologies). The two prototypes have been discussed on the basis of the characteristics mentioned above. We have shown how both the prototypes have been effectively used for the evaluation of interactive features of products, when the products are not yet available. The tests performed using these kinds of VPs can give useful feedback about the users' acceptance of new products aesthetic and interaction features, about the products ergonomics and usability, and about users' preferences and expectations about the product.

One of the main issues of product virtualization concerns the development of prototypes that are sufficiently faithful and effective for providing information about the acceptance of the future product by users, which can be used to feed back the product design, early in the concept phase. It is still difficult to create a high-fidelity VR experience, generally speaking, due to large technical limitations on processing power, image resolution and interaction modalities. For example, despite the fact that several research groups are working on the development of haptic and tactile devices, the technology is still far from being effectively used, especially for the simulation of the haptic interaction with real products. Nevertheless, the technology developers hope that such limitations will be overcome as technologies become more powerful and cost-effective over time.

Acknowledgement The author would like to thank Giandomenico Caruso and Francesco Ferrise, and all members of the KAEMaRT Group (www.kaemart.it) for their contribution to the development of this research.

References

1. Roozenburg N, Eekels J (1995) Product design: fundamentals and methods. John Wiley Sons, Chichester
2. Ulrich KT, Eppinger SD (2008) Product design and development. McGraw Hill, New York
3. Cross N (2001) Engineering design methods. John Wiley Sons, Chichester
4. Rodgers P, Patterson A, Wilson D (1996) A formal method for assessing product performance at the conceptual stage of the design process. In: Gero JS (ed) Advances in formal design methods for CAD. Chapman and Hall Inc., London
5. Virzi RA, Sokolov JL, Karis D (1996) Usability problem identification using both low- and high-fidelity properties. Proceeding SIGCHI conference on human factors in computing systems: common ground, Vancouver, BC, Canada
6. Wang GG (2002) Definition and review of virtual prototyping. JCISE 2(3):232–236
7. Schmalstieg D, Fuhrmann A, Hesina G, Szalavari Z, Encarnacao LM, Gervautz M, Purgathofer W (2002) The studierstube augmented reality project. Presence: Teleoperators and Virtual Environments 11(1):33–54
8. Bimber O, Raskar R (2005) Spatial augmented reality: merging real and virtual worlds. A K Peters, Ltd, Wellesley, Massachusetts
9. Yang K (2007) Voice of the customer: capture and analysis. McGraw-Hill Professional, New York
10. Ficalora JP, Cohen L (2009) Quality function deployment and Six sigma–A QFD Handbook, 2nd edn. Prentice Hall, Englewood Cliffs NJ
11. Nagamachi M (2008) Perspectives and the new trend of Kansei/affective engineering. The TQM J 20(4):290–298

12. Nagamachi M (1999) Kansei Engineering; the Implication and Applications to Product Development, Proceeding IEEE International Conference on Systems, Man, and Cybernetics 1999, Tokyo, Japan
13. Pham DT, Dimov SS (2001) Rapid manufacturing–the technologies and applications of rapid prototyping and rapid tooling. Springer, London
14. Malone E, Lipson H (2007) Fab@Home: the personal desktop fabricator kit. Rapid Prototyp J 13(4):245–255
15. Sells E, Smith Z, Bailard S, Bowyer A, Olliver V (2009) RepRap: the replicating rapid prototype: maximizing customability by breeding the means of production. In: Piller FT, Tseng MM (eds) Handbook of research on mass customization and personalization. World Scientific Publishing Company, Singapore
16. Horiuchi S, Kanai S, Kishinami T, Hosoda S, Ohshima Y, Shiroma Y (2005) Low-cost and rapid prototyping of UI-embedded mock-ups using RFID and its application to usability testing. Proceeding. HCII 2005 Conference, Las Vegas
17. Aoyama H, Nordgren A, Yamaguchi H, Komatsu Y, Ohno M (2007) Digital style design systems from concept to sophisticated shape. IJIDeM 1:55–65
18. Jayaram S, Vance J, Gadh R, Jayaram U, Srinivasan H (2001) Assessment of VR technology and its applications to engineering problems. J Comput Inf Sci Eng 1(1):72
19. Vo DM, Vance JM, Marasinghe MG (2009) Assessment of haptics-based interaction for assembly tasks in virtual reality. Proceedng World Haptics 2009, Salt Lake City, UT, USA
20. Azuma R, Baillot Y, Behringer R, Feiner S, Julier S, MacIntyre B (2001) Recent advances in augmented reality. IEEE Comput Graphics Appl 21(6):34–47
21. Burdea G, Coiffet P (2003) Virtual reality technology. Wiley, New Jersey
22. Craig AB, Sherman WR, Will JD (2009) Developing virtual reality applications–foundations of effective design. Morgan Kaufmann Publishers, Elsevier, Burlington
23. Hayward V, Astley OR, Cruz-Hernandez M, Grant D, Robles-De-La-Torre G (2004) Haptic interfaces and devices. Sens Rev 24(1):19–29
24. SenSable. http://www.sensable.com. Accessed 1 Feb 2011
25. MOOG-HapticMaster. http://www.moog.com/products/haptics-robotics/. Accessed 1 Feb 2011
26. Haption. http://www.haption.com/. Accessed 1 Feb 2011
27. Bordegoni M, Cugini U (2008) The role of haptic technology in the development of aesthetic driven products. J Comput Inf Sci Eng 8(4):1–10. doi:10.1115/1.2988383
28. Cugini U, Bordegoni M (2007) Touch and design: novel haptic interfaces for the generation of high quality surfaces for industrial design. Visual Comput, 23(4):233–246
29. Bordegoni M, Ferrise F, Covarrubias M, Antolini M (2010) Haptic and sound interface for shape rendering. Presence: Teleoperators and Virtual Environments 19(4):341–363
30. Wu J, Yu G, Wang D, Zhang Y, Wang CCL (2009) Voxel-based interactive haptic simulation of dental drilling, Proceeding ASME IDETC/CIE 2009 Conference, San Diego (CA) USA
31. Bordegoni M, Colombo G, Formentini L (2006) Haptic technologies for the conceptual and validation phases of product design. Comput Graphics 30(3):377–390
32. Human factors in auditory warnings (1998) Stanton NA, Edworthy J (eds) Ashgate Publishing
33. Lemaitre G, Houix O, Visell Y, Franinovic K, Misdariis N, Susini P (2009) Toward the design and evaluation of continuous sound in tangible interfaces: the spinotron. Int J Hum-Comput St 67(11):976–993
34. Rocchesso D, Polotti P, Delle Monache S (2009) Designing continuous sonic interaction. Int J Design 3(3):13–25
35. Delle Monache S, Polotti P, Rocchesso D (2010) A toolkit for explorations in sonic interaction design. Proceeding 5th Audio Mostly conference: a conference on interaction with sound, Pitea, Sweden
36. Farnell A (2010) Designing sound. The MIT Press, Cambridge
37. O'Brien JF. Shen C, Gatchalian CM (2002) Synthesizing sounds from rigid-body simulations. Proceeding ACM SIGGRAPH/Eurographics symposium on Computer Animation (SCA '02)

38. Bordegoni M, Ferrise F, Shelley S, Alonso M, Hermes D (2008) Sound and tangible interface for shape evaluation and modification. Proceeding HAVE 2008–IEEE International Workshop on Haptic Audio Visual Environments and their Applications, Ottawa, Canada
39. Bordegoni M, Ferrise F, Lizaranzu J (2010) Multimodal interaction with a household appliance based on haptic, audio and visualization. Proceeding IDMME–Virtual Concept 2010, Bordeaux, France
40. Haller M, Billinghurst M, Thomas B (2007) Emerging technologies of augmented reality–interfaces and design. Idea Group Publishing, London
41. Azuma R, Baillot Y, Behringer R, Feiner S, Julier S, MacIntyre B (2001) Recent advances in augmented reality. IEEE Comput Graphics Appl 21(6):34–47
42. Milgram P, Takemura H, Utsumi A, Kishino F (1994) Augmented reality: a class of displays on the reality-virtuality continuum. Proceeding telemanipulator and telepresence technologies, Boston, MA, USA
43. Neumann U, Cho Y (1996) A self-tracking augmented reality system. Proceeding ACM Symposium virtual reality software and technology, Hong Kong
44. Caruso G, Cugini U (2009) Augmented reality video see-through HMD oriented to product design assessment. Proceeding 3rd International Conference on virtual and mixed reality, San Diego, USA
45. Cakmakci O, Rolland J (2006) Head-worn displays: a review. J Disp Technol 2(3):199–216
46. Bordegoni M, Cugini U, Caruso G, Polistina S (2009) Mixed prototyping for product assessment: a reference framework. Int J Int Des Man 3(3):177–187
47. Dai F (1998) Virtual Reality for Industrial Applications. Springer, New York
48. Nam TJ, Lee W (2003) Integrating hardware and software: augmented reality based prototyping method for digital products. Proceeding CHI, Fort Lauderdale, FL, USA
49. Verlinden J, Horvath I (2006) Framework for testing and validating interactive augmented prototyping as a design means in industrial Practice. Proceeding virtual concept, Playa del Carmen, Mexico
50. Kimishima Y (2006) Development of evaluation system for style design using mixed reality technology. Proceeding ASME-IDETC/CIE, Philadelphia, PA, USA
51. Kanai S, Horiuchi S, Kikuta Y, Yokoyama A, Shiroma Y (2007) An integrated environment for testing and assessing the usability of information appliances using digital and physical mock-ups. Proceeding. HCI 2007, Beijing, P.R. China
52. Bordegoni M, Polistina S, Carulli M (2010) Mixed reality prototyping for handheld products testing, Proceeding IDMME–virtual concept 2010, Bordeaux, France
53. Ferrise F, Bordegoni M, and Lizaranzu J (2010) Product design review application based on a vision-sound-haptic interface. In: Nordahl R, Serafin S, Fontana F, Brewster S (eds.) Haptic and audio interaction design, Springer Berlin/Heidelberg, 6306:169–178
54. Bordegoni M, Ferrise F, Lizaranzu J (2011) Use of interactive virtual prototypes to define product design specifications: a pilot study on consumer products. Proceeding IEEE-ISVRI 2011, Singapore
55. Caruso G and Cugini U (2009) Augmented reality video see-through HMD oriented to product design assessment. In virtual and mixed reality
56. Kato H, Billinghurst M (1999) Marker tracking and HMD calibration for a video-based augmented reality conferencing system. Proceeding 2nd international workshop on augmented reality, San Francisco, CA, USA
57. Foley J, Van Dam A, Feiner SK, Hughes JF (2000) Computer graphics-principles and practice. Addison-Wesley, Boston

Chapter 8
Digital Human Models Within Product Development Process

Caterina Rizzi

Abstract This chapter focuses on the use of Digital Human Models (DHM) within the product development process enabling the designers to address and solve ergonomics and human factors along the whole product life cycle. First part of the chapter introduces the state of the art of DHM tools. It comprises a short history, which points out the evolution of virtual humans since their origins, and proposes a classification taking into account also fields of application. Second part of the chapter describes possible applications of DHM within the product development process. Case studies from different industrial contexts are presented and discussed. The first regards the use of DHM and VR techniques in automotive and farm tractor industry, the second the integration of virtual prototyping techniques with DHM to validate and study new solutions for refrigerated unit and machinery and the last one the virtual testing of artificial lower limb prosthesis. Final remarks conclude the chapter.

8.1 Introduction

During the last decade, Digital Human Models–DHM (also named Virtual Humans–VH) are becoming more and more popular in different industrial domains since their application can be various within the product development process [1–3]. Companies are, in fact, realizing that human issues are not sufficiently considered during the design, assembly, maintenance of products and processes. Therefore, virtual humans can represent a valid tool to support the design team during the product development process from the early phases to the disposal, by decreasing the

C. Rizzi (✉)
Department of Industrial Engineering, University of Bergamo,
Viale G. Marconi, 5, 24044 Dalmine, BG, Italy
e-mail: caterina.rizzi@unibg.it

M. Bordegoni and C. Rizzi (eds.), *Innovation in Product Design*,
DOI: 10.1007/978-0-85729-775-4_8, © Springer-Verlag London Limited 2011

development time, reducing the need of physical prototypes, lowering costs and improving the quality and safety of the products [1].

Several techniques and software tools have been developed and are commercially available, most of them originated from research activities. Different models of the human body or of its parts are required at different levels of detail depending on the target application. For example, ergonomics analyses of a product need a biomechanical model able to replicate the human movements; while some biomedical applications need a detailed description of anatomical district under investigation, including internal parts, such as muscles, bones or even blood vessels and so on. This means that we need to consider different techniques and tools to create the adequate human models able to reproduce key features for each domain.

In this chapter we consider systems and applications for product development process with reference to the so-called Physical Digital Human Models [2]. The first part of the chapter presents a state of the art on digital human modelling which includes a short history, proposes a classification and mentions main applicative domains; the second part describes the application of DHM within the product development process as well as some practical examples developed in various industrial contexts.

8.2 State-of-the-Art of Digital Human Modeling

Digital human modeling started in the 1960s and at present, various frameworks for virtual humans of different complexity can be found on the market or are developed in academia [1–3]. In fact, many research activities are under development with the goal of fulfilling the requirements coming from different industrial sectors. Beside commercial systems, we can find academic models developed to solve specific problems, such as the work carried out by Valentini et al. [4] for the automotive vibration comfort analysis, the human models developed by Magnenat-Thalman et al. for clothing [3], and biomechanical applications. Before introducing a classification, a brief history of human modelling is presented.

8.2.1 A Short History

The earliest applications of virtual humans go back to the late 1960s and refer to ergonomics analysis, mainly in aeronautics and automotive industries.

The first virtual manikin, named "First Man" (later Boeman) was developed for Boeing in 1959 to assess pilot accommodation in the cockpit of a Boeing 747 [3]. "First Man" evolved up to the "Fourth Man and Woman" version, the first virtual human displayed as a set of colored polygons on a raster device [5]. In 1969, Boeman was developed at Boeing Corporation. It was based on anthropometric data of a 50th percentile man, with 23 joints and links of variable length [6].

Always at Boeing Corporation, it was implemented Combiman–Computerized biomechanical man-model (later CrewChief) by which a manikin could be generated using anthropometric data from six military databases. It permitted Boeing engineers to test how the pilot could reach objects in a cockpit, taking into account the effect of clothing and harnesses.

Similarly, in the automotive domain, various human models have been designed to support the car design process since the 1970s. An example is Cyberman (Cybernetic man-model) developed by Chrysler Corporation for ergonomic analysis inside and outside a car. The manikin was made of 15 segments and the position of the observer was predefined [3].

SAMMIE (System for Aiding Man-Machine Interaction and Evaluation) was designed in 1972 and was one of the first computer program oriented to ergonomic design and analysis [7]. It consisted of a parameterized human model made of 21 rigid links and 17 joints, a well developed vision system and manipulation of complex objects.

Later on, a large number of human modeling tools for a wide range of applications have been developed exhibiting increased capabilities and functionality, thanks also to the increasing power of computers.

In the meantime, since the beginning of 1980s, several research groups and companies started to produce movies using virtual humans [3]. Among the others, we can quote "Virtual Marilyn" by N. Magnenat-Thalmann and D. Thalmann in 1987, "Tin Toy" in 1988 or the short movie "Geri's Game" from Pixar.

Many research groups have been working on virtual humans with applications in different contexts, from automotive, to movies and video-games, and various commercial systems are available. D. Chaffin at the University of Michigan, N. Badler at the University of Pennsylvania, N. Magnenat-Thalmann and D. Thalmann contributed considerably to the foundation of today's human modeling [2]. As an example, in the mid 1980s N. Badler and his research team developed Jack, a computer manikin software, now part of Siemens PLM Solutions suite.

In addition, military authorities continuously funded researches for the development of more sophisticated digital humans [2]. An example is Santos, a virtual soldier, developed by the Virtual Soldier Research Group at Iowa University since 2003. It acts as an independent agent and, apart from common features, it includes realistic skin deformation and contracting muscles.

8.2.2 A Classification

As previously mentioned, many systems are available to create and animate virtual humans. They can be used to show and analyze how humans should act in various situations and execute required tasks, but also to predict the impact of their actions (in terms of postures and movements) on their muscoloskeletal apparatus.

By analyzing the state of the art and the types of applications, the digital human models have been grouped into four main categories:

Fig. 8.1 Garment simulation
on a standard manikin and
customer's virtual body

- **Digital humans/actors for entertainment** [3, 8–12]. Virtual actors are used to populate scenes for movies and videogames production. An example is Poser [8] by Smith Micrographics, which is specialized in human body modeling and rendering. Tools such as Massive [10], CrowdIT [11] and Di-Guy [12] are virtual crowd simulators; typically they are used to create virtual scenes populated with huge numbers of individuals (autonomous agents), for example to represent platoons, fighters or civilians. They can also be used to simulate emergency situations and for training purposes.
- **Virtual Manikins for Clothing** [3, 13–21]. They are used to create virtual catwalks, catalogues, try-on show rooms and to design garments. The applications are various and imply the use of virtual mannequins with different levels of complexity according to the goal, from virtual catwalk to garment design. In addition, research activities have been done or are under development to generate virtual mannequin/human directly from body scanner or similar [15, 21] and/or to extract customer's anthropometric measures for custom-fit garment design. Modaris, Vsticher and 3-D Runway are some of the commercial CAD clothing systems that use 3-D virtual mannequins. As an example, Fig. 8.1 shows a standard virtual manikin and a virtual body of a customer acquired with a body scanner, both wearing the same garment.
- **Virtual Humans for Ergonomic Analysis** [2, 22–38]. These digital humans, also called computer manikins [2], are three-dimensional models of the human body designed to assess compatibility and usability of products, machinery or workplaces with the human user or operator. It is possible to create manikin, both male and female, of different sizes using anthropometric databases based on recent military (e.g., US Army) and civilian anthropometric surveys. A virtual manikin is a kinematic chain composed by a number of rigid links connected by joints; each virtual joint reproduces the degrees of freedom (d.o.f.) of the physical one, with limits corresponding to the allowable motion of a human being. Some software packages available on the market are Jack [23, 24], Ramsis [25], Safework [26], HumanCAD [27] and BHMS [28]. These systems permit us to define complex scenes, analyze postures, simulate tasks and optimize working environments and are used by a broad range of companies,

Fig. 8.2 Virtual human at
the driver's seat (image
courtesy [30])

especially in automotive and aeronautics domains. For example, Ramsis is
widely used by companies such as Audi, Volkswagen, Ford and Honda [25],
Jack by John Deere and BAe Systems [24]. In the automotive domain the
applications range from ergonomic design of car interior (driver's seat, dashboard instruments and controls, etc.) [29, 30], to comfort analysis [4], car
ingress/egress simulation [31], and manufacturing processes [32, 33]. Figure 8.2
shows the use of Jack mannequin for interior car ergonomic analysis. Similarly
in aeronautics, virtual mannequins have been used to design more efficiently the
airplane interiors, such as the cockpit [34–37]. For example, from 1987 Boeing
Corp. developed an ad hoc tool, named BHMS. Virtual humans have been also
used to study loading and unloading systems of luggage in airports [38].

- **Detailed Biomechanical models** [39–46]. These models are more complex than
  the previous ones; in fact, they are characterized by a complete muscoloskeletal
  model. They are used to compute biomechanical system kinematics using
  forward and inverse kinematics algorithms. As for to the previous category,
  applications are various and include ergonomics analysis, gait analysis, study
  and simulation of the human body during movements, by considering
  also people with different pathological conditions [39] or disabilities, the
  rehabilitation and the validation of new devices [40, 41], etc. Tools, such as
  LifeMOD [42], SIMM [43], MADYMO [44] and Santos [45, 46] belong to this
  group. For example the Santos human model includes anatomy, biomechanics,
  physiology, and intelligence in real-time [46]. Its main feature is the *predictive
  dynamics,* i.e., Santos "predicts how people strike various poses and how they
  move and act" [46].

Most of these systems implement or incorporate analysis tools such as RULA
(Rapid Upper Limb Analysis) to investigate work-related upper limb disorders,
NIOSH (U.S. National Institute for Occupational Safety and Health) lifting
equations to evaluate lifting and carrying tasks, and OWAS (Owako Working
Posture Analysis System) to analyze postures during work.

From the previous analysis one can derive that digital humans and related tools
can be used in various areas/applicative domains according to the specific needs;
some examples are the following [2, 3]:

- Engineering design and manufacturing
- Automotive
- Aeronautics
- Transport (e.g., track, bus and train)
- Architecture & Civil Engineering
- Textile and Clothing industry
- Bioengineering and Medicine (e.g., rehabilitation and surgery)
- Videogames and Movies
- Sport science
- Education & Training
- Museums
- …

In the next sections we main focus the attention on the use of digital human modeling within the product development process and some case studies will be described to show human models potential.

8.3 Digital Human Modeling in Product Development Process

During last decades computer-based systems, such as CAD and CAE (Computer-Aided Engineering) packages, have been widely and commonly used within the product development process since its early stage up to the product production and delivering. However, designers have to take into account that the product will be used by human beings, acting as workers, users or consumers, which interact with the product in different ways according to their needs. This requires (1) the development of products centered on human beings and suitable to the widest range of population characterized by different sizes, genders, ages, preferences and abilities [2], and (2) the adoption of strategies and tools that permit to consider ergonomics aspects since the conceptual design stage. In such a context, DHM tools can enable the designers to address and solve ergonomics and human factors along the whole product life cycle. They can be used for product design, crash testing, product virtual testing, workplace design, and maintenance allowing a fast redesign and reducing the need and realization of costly physical prototypes, especially for those applications dealing with hazardous or inaccessible environments. Therefore, the use DHM tools is not limited to the early stage of product life cycle but it can be profitably also in the following phases up to service and maintenance applications [2].

They also promote the respect of standards (e.g., European Standard UNI-EN 1005-4:2005 and UNI-EN 1005-5:2007), which specify requirements for workers' postures and movements in order to reduce health risks.

Considering product design phase, DHM enables designers to face several ergonomics issues, such as [47]:

- *Comfort and posture prediction* for envisaged groups of users; for example if the product dimensions can fit the various users' sizes.
- *Task evaluation and safety* to analyze if a task can be ergonomically executed and ensure workers' safety in order to avoid injuries or prevent muscoloskeletal diseases; in fact, many workers perform repetitive tasks that can cause pain and fatigue.
- *Visibility* to check if the users can opportunely see the product (machinery, equipment, vehicles, etc.) or its functional sub-systems (e.g., car control panel) when using or manipulating it in different working conditions.
- *Reach and grasp* to verify if the disposal of devices (e.g., buttons, sliders and shelves) or product configuration permits all users' groups to easily access and manipulate them, for example during maintenance operations.
- *Multi-person interaction* to analyze if and how multiple users can interact among them and with the product.

In addition, DHM can be integrated either with VR (Virtual Reality) systems to improve the level of interaction and realism within the virtual environment or with Mocap (Motion Capture) equipment to drive the virtual human and facilitate the evaluation of comfort and prediction of injuries that could rise when executing a task.

In the next three sections, the author presents applicative examples of DHM in different industrial context: the first section describes the use of DHM and VR techniques in automotive and farm tractor industry, the second the integration of virtual prototyping techniques with DHM to validate and study new solutions for refrigerated unit and machinery and the last the virtual testing of artificial lower limb prosthesis.

8.3.1 DHM and VR Techniques in Automotive and Farm Tractor Industry

Visibility is an ergonomic issue related to a wide variety of products, machinery and workplace. In the following, two examples related to the use of DHM tools to analyze visibility are presented. In both cases, Jack human model has been adopted.

The first application concerns car interior design described in [30]. In this case, visibility can be subdivided into sub-problems: internal visibility of instruments and controls and external visibility through the windscreen, windows and mirrors.

Figure 8.3 shows the model of the car interior used to perform the simulation.

The virtual driver is in a normal driving posture (Fig. 8.2) and the visibility analyses, in particular of the dashboard behind the wheel, have been performed varying manikin size (both male and female from 50 to 95th percentile) and the positions of the seat and of the steering wheel. As an example, Fig. 8.4 shows a sequence of images for 50th percentile manikin and for the driver's seat at the

Fig. 8.3 3-D digital model
of car interior (image
courtesy [30])

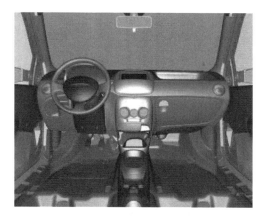

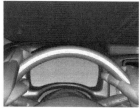  

Fig. 8.4 Instruments panel visibility for a 50th percentile: seat at foremost position and steering
at the three different positions

foremost position and steering wheel at three different positions (low, medium and
high). One can note that, even if in some positions of the steering wheel the
dashboard visibility is adequate, the driver's posture is not correct and the seat is
too close to the wheel limiting also the external visibility.

Simulations for external visibility analysis was more articulated and took into
account two recurrent situations: car motionless with objects and people moving
around the car itself and car moving along predefined paths. The analysis mainly
concerned problems due to the presence of some parts of the car that could hide
objects, e.g., traffic lights and pedestrians, and therefore limit the visibility.
Figure 8.5 shows the virtual environment representing the second situation:
pedestrian is moving around the car and some markers define the points of the path
and the simulation sequence. Also in this case a series of simulations have been
carried out varying the manikin size, and the seat position. The figure portrays a
sequence of images captured during such a test; the visibility lacks due to the door
pillars are clear. The head motion is essential since in this case the visibility cannot
be simply simulated with virtual cameras and the complex kinematic chain supporting the cameras plays a relevant role. Head motion must be programed
specifying joint law [30].

A similar problem can be found during the design of farm tractors for vineyard
and orchard [48]. This type of tractor is equipped with different tools to

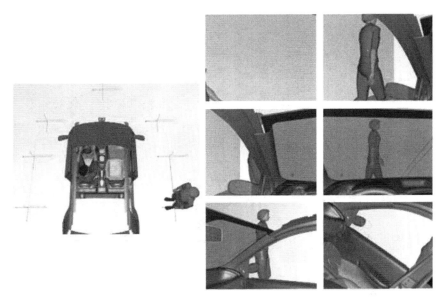

Fig. 8.5 Virtual environment and walking pedestrian as seen by the driver through car windows (image courtesy [30])

accomplish to various tasks (e.g. harvesting and fruit picking); therefore, a specific ergonomic aspect consists in the external visibility of mounted equipment, such as a loading shovel, through the front window and the high-visibility roof. Figure 8.6 shows the simplified model of the tractor, the manikin and three different positions of the loading shovel (identified by different colors). In addition, a vertical grid has been considered to evaluate quantitatively the visible area of the equipment. The manikin posture has been identified according to SAE (Society of Automotive Engineers) standard, which provides for each joint of the virtual human the range of acceptable angles to assure a comfortable position. The wheel has been considered fixed while the seat can move horizontally (X-axis, −70 mm, +80 mm) and vertically (Z-Axis, −50 mm, +50 mm). Simulations have been planned in order to analyze:

1. Visibility of the loading shovel through the front window.
2. Visibility of the loading shovel through the high-visibility roof.
3. Visibility through the front window of the loading shovel located at the lowest position (ground).

The simulations have been carried out varying the manikin size and seat position horizontally and vertically. For simulation (1), the optimal position for each manikin was first identified and, then, the positions of the loading shovel when it begins to be visible, to be hidden by the shade curtain or totally disappears. For instance, Fig. 8.6 shows the comparison of results obtained for 5th, 50th and

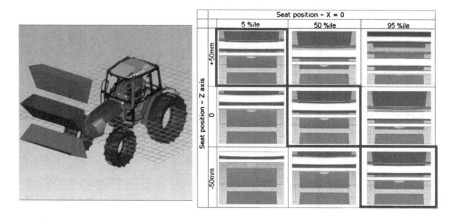

Fig. 8.6 Virtual prototype of the tractor (*left*) and frontal visibility varying manikin sizes and seat position along Z-axis (*right*)

95th percentiles varying the seat position only vertically (along Z-axis) and highlights best configuration for each manikin.

Analyses like those described, allow the designer to evaluate ergonomic aspects such as posture or visibility but they do not permit "fine" interaction and, for example, the evaluation of the roughness of surfaces or the deformability of a part [49].

As mentioned before, Virtual Reality (VR) can help to add realism and improve ergonomic analysis and human modeling tools play an important role since they permit to reproduce the movement of the real user in the virtual scene. For example, it is possible to carry out ergonomics analysis of a car dashboard, in a virtual environment and simulate traditional interaction between the tester and the product using haptic devices. Typically, the tester is immersed in a virtual world (a 3-D representation of the product, the digital human and other objects) where s/he can manipulate the control devices (knob, slider and button) and receive haptic feedback.

Colombo et al. in [49] presented an ergonomic workstation that integrates VR and haptics to carry out virtual ergonomic tests based on visual and tactile interaction of car control panels. Figure 8.7 shows the architecture of the workstation and its main components, both hardware and software. It integrates commercial components (displaying system, tracking system, software) and "ad hoc" developed ones (haptic devices).

The virtual environment is created by a virtual reality engine, which manages and visualizes the scene and the tester; the avatar reproducing postures and movements of the tester is essential to ensure a good level of immersion and realism. The virtual scene is visualized on an HMD (Head Mounted Display) and on a wide screen (but also a computer monitor can be used). The tracking system (in this case the VICON system equipped with four high performances

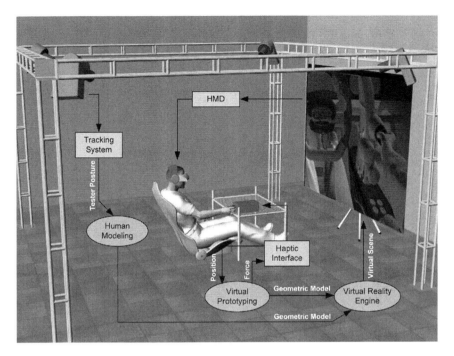

Fig. 8.7 Proposed workstation for ergonomic analysis of car control panel (image courtesy [49])

infrared cameras) detects the postures and the motion of the tester and transmits them to Jack software to update the human model. When the tester touches and modifies the position of a control, the haptic device communicates the new position to the virtual reality engine that updates the scene; the force-feedback is instead managed entirely by the haptic device. The complete virtual scene (product with control board, human avatar and other objects) is assembled and managed by Jack. Jack also calculates the tester's point of view on the scene and modifies the scene accordingly. This image is then transmitted to the HMD and to a projector in order to represent the virtual scene to the tester and to the audience. Figure 8.8 portrays an image of the simulation tests.

In such a way, the tester can rapidly evaluate different design solutions thanks to the haptic devices that permit to simulate different types of controls (knobs, buttons and sliders).

A similar application is described in [50] where the authors present a Mixed-Reality environment to validate the redesign of a farm tractor cabin. The environment, based on haptic interface, permits to test and validate ergonomically different command layouts.

Both experience permitted to demonstrate the effective potential of DHM tools for ergonomics analysis and their applicability in industrial contexts to improve product ergonomics and to compare different design solutions.

8.3.2 Integration of VP and DHM for Ergonomics Analysis of Refrigerated Cabinets and Machinery

The case-study of refrigerated cabinet considered virtual human modeling as a tool to integrate virtual prototyping and carry out ergonomic analysis during the development process of refrigerated cabinets, commonly used in the supermarket to display packed food [51]. The design of such a machinery should accommodate the full range of users during its life cycle; i.e., from workers like maintenance men to the final users. In this case, different ergonomic issues have been considered: product visibility and reach, posture comfort and safety and multi-user interactions. Three sets of simulation campaign have been considered in order to:

1. Validate the refrigerator units design from customers' and supermarket operators' point of view taking into account different ergonomics issues. For example, customers should easily access food packages, while supermarket operators have to execute repetitive tasks (e.g., loading a shelf) with postures and movements as safely as possible without causing musculoskeletal disorders, health risks and respecting hygienic constraints.
2. Design auxiliary equipment for shelves loading and ergonomically validate proposed solutions from the operator's point of view using methods such as NIOSH, OWAS and Lower Back analysis.
3. Study a solution for electric equipment maintenance suitable for different types of refrigerated units. The design solution should allow technicians to access

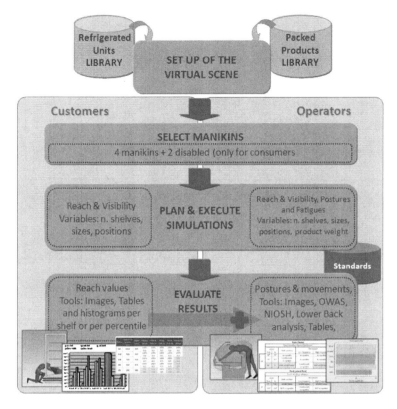

Fig. 8.9 Roadmap for ergonomics analysis with digital human models

parts, manipulate tools necessary for the task [51] and ensure part removal and replacement, visibility during the task execution and technician's safety.

The basic idea was to define a step-by-step roadmap the designer can adopt to evaluate the ergonomic aspects and analyze alternative configurations of the refrigerated units since the early stage of the product development. Figure 8.9 portrays the roadmap based on the use of 3-D parametric models representing product archetypes and virtual humans.

The commercial software pacakge Jack has been used to model the virtual human and perform the simulations. Concerning the set up of the virtual scene, two libraries have been developed: one for the refrigerated units and another one for the packed food. The first includes 3-D parametric models, which are archetypes of the units. Each 3-D model is a simplified representation of the product where components specifically involved in the ergonomics analysis have been parameterized to be easily modifiable, e.g., shelves sizes (length and width), positions (height from ground) and number. The second library includes a set of 3-D parametric models representing a wide range of products easily adaptable to the specific refrigerated unit. To take into account diversity of people size (consumers

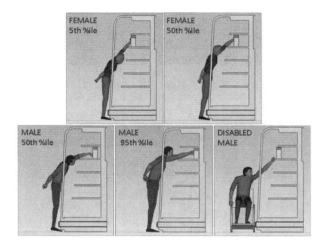

Fig. 8.10 Comparison of manikin postures and reach & grasp for the 2nd shelf from the top

and operators) 6 manikins were considered: 2 females (5th and 50th percentiles), two males (50th and 95th percentiles) and only for consumers, two manikins (female and male 50th percentile) sitting on a wheel chair representing people with reduced mobility.

For the first campaign, simulations have been performed for five types of refrigerated units included in the library, by varying parameters such as shelves positions and number.

Regarding the customers, results were quantified by calculating the farthest distance reachable from the frontal edge of each shelf and with respect to the center of the hand of the manikin in order to evaluate product and shelves accessibility. Quantitative data were organized in tables and histograms allowing the designer to evaluate rapidly not only the specific machinery but also to compare different types of refrigerated units with respect to packed food accessibility and expository space.

As an example, Fig. 8.10 shows the results related to the second shelf of a vertical refrigerated unit with six shelves; while Table 8.1 summarizes quantified values. One can note that the 4th shelf can be accessible by all considered population, the 3rd, 4th and 5th have values higher than 60%, while the 1st and 2nd are the most critical, especially the first one is accessible only by the 95th percentile. The disabled man cannot reach the 1st and 2nd shelves, while the other shelves are partially accessible.

Apart from product visibility and reach, postures and comfort have been analyzed for the customer's operator using as reference international standards that specifies the requirements for postures and movements in order to reduce health risks and methods such as OWAS and NIOSH. In fact, operators are exposed to repetitive tasks that can cause pain and fatigue; this means that the engineers should design machinery reducing as much as possible painful and tiring postures

Table 8.1 Reaching values for the vertical unit

Shelves	Height from ground [mm]	Depth [mm]	FEMALE 5th %ile	FEMALE 50th %ile	MALE 50th %ile	MALE 95th %ile	DISABLED MALE
1st	1812	414	0% 0	2% 10	24% 100	56% 230	0% 0
2nd	1537	463	15% 70	37% 170	55% 260	100% 463	0% 0
3rd	1262	514	62% 320	82% 420	82% 420	100% 514	30% 150
4th	987	565	100% 565	100% 565	90% 510	100% 565	58% 300
5th	712	619	61% 380	70% 430	78% 480	78% 480	53% 330
6th	296	787	64% 500	64% 500	67% 530	66% 520	33% 260

Table 8.2 Operator's results

Hisghest shelf				Female 5%	Female 50%	Male 50%	Male 95%
Product reach			%	40	50	100	100
Posture			OWAS Code	1	1	3	2
F A T I G U E	1 Kg	Lower Back	L4L5 (Nm)	30	20	55	95
			Spinal Forces (N)	800	1500	1200	2050
			Muscle tension (N)	250	100	400	750
			Static force	150	150	275	390
		Recover time	Cycles	25	27	29	28
			RT (sec)	8,19	4,03	0,12	2,17
	3 Kg	Lower Back	L4L5 (Nm)	45	25	70	110
			Spinal Forces (N)	1000	700	1800	220
			Muscle tension (N)	270	240	500	770
			Static force	150	420	280	390
		Recover time	Cycles	25	27	29	27
			RT (sec)	10,32	5,07	1,23	2,75
NIOSH			RWL	3,88	3,65	3,37	3,65
			LI	0,39	0,41	0,45	0,41
			CLI	1,39	1,48	1,6	1,47

Female 50%

and movements, and thus health risks. Table 8.2 reports the results obtained for the highest shelf.

The results from the previous campaign have highlighted the need of auxiliary equipments specifically designed to load shelves.

During the second simulation campaign, three auxiliary equipments (lift carriages) have been compared (Fig. 8.11).

Even if the new solutions improve product reach and avoid some risky postures (see figure included in Table 8.2), still some problems remain. For example,

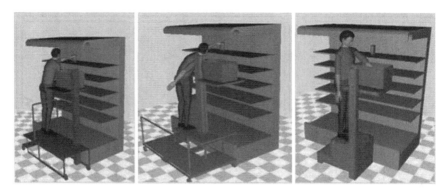

Fig. 8.11 Auxiliary equipments

uncomfortable postures assumed to load the lowest shelf are not solved and the presence of a shelf in the auxiliary equipment to support products container, in some case, hinders/limits operators' movements. Therefore, new design solutions have been proposed and tested as well.

Human modeling and simulation results permitted to validate alternative design solutions evaluating posture comfort and risk of musculoskeletal diseases, thus operators' safety, before physical implementing them. In addition, thanks to the library of the 3-D parametric model, it was quite easy to modify meaningful dimensions of the unit or shelves (sizes, location and number) to find better ergonomic performances.

The same methodology has been adopted to study a solution for electric equipment maintenance appropriate for all types of units. In fact, in some units, the equipment was located over the roof; while in others it was mounted close to the floor. In the first case, it was necessary to use a stair and, often, not all parts were easily reachable and visible by the maintenance man; in the second case s/he had to assume uncomfortable posture to see, access and replace parts (Fig. 8.12–As-Is). The solution proposed was a connector to create a link with the electric equipment thus creating an easy accessible checkpoint. The simulation carried out for the different units permitted to identify the correct position for the connector link as shown in Fig. 8.12 (To-Be).

Ergonomic analysis of work task is a typical application of human models and may concern several types of products often manufactured by SMEs (Small-Medium Enterprises). The case-study reported in [30] is related to a workstation equipped with an automatic riveting system to assembly household products produced by a small manufacturer of machine tools. There are several risks for the operators: assumptions of uncorrected postures for long periods of time, inadequacy of the sitting positions, reiterated movements of legs, frequent flexion and torsion of the bust, manual movements of loads with slanting bust or in torsion. Human modeling has been used to analyze the level of comfort for the operator by

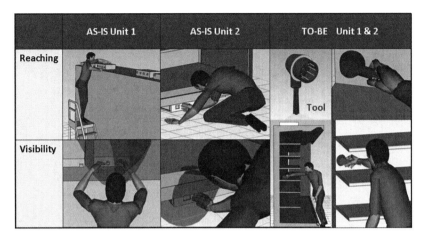

Fig. 8.12 As-Is and To-Be configuration for electric equipment

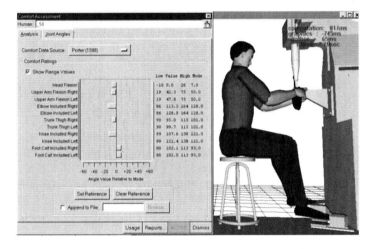

Fig. 8.13 Acceptable values (green) for joints angles (image courtesy [30])

varying the virtual workstation configuration (e.g., stool position and height or treadle drive orientation) and, therefore, to evaluate different working conditions and postures. The angles assumed by the human joints for each adjustment of the workstation have been analyzed to identify acceptable configuration. Figure 8.13 portrays the acceptable values of the joints angles for a simulation carried out with a 50th percentile virtual human and a given configuration of the work station.

In this case, RULA (Rapid Upper Limb Assessment) and NIOSH methods have been considered. The results obtained allowed the manufacturer to define configurations ensuring a good level of comfort for the operator.

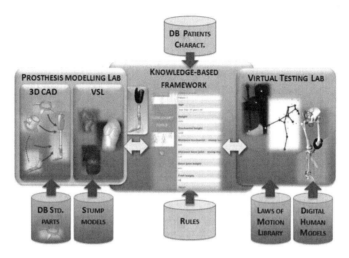

Fig. 8.14 High-level architecture of the framework

8.3.3 *Virtual Testing of Lower Limb Prosthesis*

Custom-fit products, such as made-to-measure garments, shoes, helmets, prosthesis, represent a challenge for design methods. Among them, lower limb prostheses are characterized by a high level of interaction with the human body and are designed around the patient or his/her anatomical districts, usually following a hand made development process. Therefore, the human body plays a central role and virtual human modeling can be a meaningful tool to improve the product development process, increase product functionality and patient's quality of life. In this context, research activities are under development to implement a new design framework totally based on the integration of CAE and human modeling tools [52]. In this case, more detailed human models or of its parts are necessary. For example, virtual modeling of a socket and the analysis of socket-stump interaction with Finite Element tools require a detailed digital model of the residual limb; while a virtual avatar can be used for the gait analysis of the patient wearing the prosthesis.

Figure 8.14 shows an example of a new framework for prosthesis design where two main virtual labs can be distinguished: Prosthesis modeling lab to model the prosthesis components around the stump digital model and the Virtual Testing lab to develop a complete amputee's digital model, a digital avatar, onto which executing the prosthesis set up and evaluating its functionality through postures and movements simulation.

For the virtual testing, a detailed biomechanical model is necessary, thus it requires the use of tools such as Santos, Anybody or LifeMod. In particular, for this application, LifeMOD, a biomechanical simulation package based on MSC ADAMS solver, has been adopted to create the patient's digital avatar. It permits

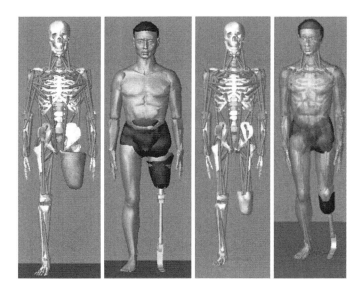

Fig. 8.15 Patients' avatar wearing the prosthesis: transfemoral (*left*) and transtibial (*right*)

to create a detailed biomechanical model of a human body using rigid links connected through joints to simulate the skeleton and flexible elements to represent soft tissues (muscle tissue, tendons and ligaments).

In order to characterize the amputee's avatar wearing the prosthesis the following data are necessary: the patient's and stump anthropometric measures, the digital models of the residual limb and of the assembled prosthesis, which correspond to three different levels of the avatar characterization. The patient's and stump anthropometric measures are essential respectively to properly size the avatar and to position and link the prosthesis model to the avatar. Figure 8.15 portrays either a transfemoral (above-knee amputation) or a transtibial (below knee) avatar wearing the prosthesis.

The avatar can be used to simulate two typical activities traditionally verified by the orthopedic technicians, such as the patient walking and sitting on a chair. LifeMOD makes available library of laws of motion acquired with MOCAP (Motion Capture) equipment. However, in this case, it is necessary to build up a customized library acquiring motion laws of amputee's joints both transtibial and transfemoral during typical daily activities: walking on flat floor, on a slope, and overcoming stairs using a Mocap system like VICON or even a low cost markerless equipment. Figure 8.16 shows first tests carried out to simulate transfemoral amputee walking.

The new product development process based on human digital models can improve the prosthesis design by speeding up the technician tasks. It is expected to permit us to reduce the number of prototypes and to lower the psychological impact on the life of the patient. In fact, a computer-aided approach allows

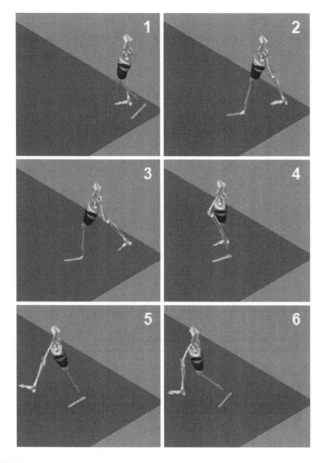

Fig. 8.16 Walking simulation of a transfemoral amputee

carrying out in a virtual way several tests of the traditional socket development process that are very bothering for amputees.

8.4 Discussion and Conclusions

This chapter presents an overview of Digital Human Modeling tools and example of applications within industrial contexts, which differ for sizes, resources and competences in using computer-aided tools to support design activities. Therefore, their industrial use is not limited to big companies but they can be also introduced into small and medium enterprises and applied to a wide range of products, such as electric appliances, furniture, boats, trains, machine tools and so on. In the product design the human factor plays a central role and in some cases products are specifically designed around the human body or part of it. This is the case of custom-fit products. This means that products should satisfy the needs of human beings that differ in several aspects, from sizes to preferences and ways of inter-action with the product.

Today's tools for human modeling offer a comprehensive and adequate set of functions and permit to study and solve several ergonomics problems the designer may face through the product life cycle reducing development times and the need of costly physical prototype.

The case studies described permitted to show the potential of DHM.

First, it has been demonstrated that it is an important tool to improve virtual prototyping functionalities and, above all, to increase ergonomics and safety of products. Range of analysis can vary from simulation of product visibility and reach to the analysis of postures evaluating comfort and safety using methods and tools, such as RULA, NIOSH and OWAS. This should also allow companies to promote and comply with of national and international standards that regulate workers' safety. However, even if many tasks performed by a human being can be reproduced, one must have in mind that, even with complex human model, human beings cannot be fully duplicated. In addition, there is still the need to identify methodological approaches and guidelines for a correct and efficient use within the product development process in a specific industrial context. Different types of analysis can be performed, but sometimes results are difficult to learn, understand and use [2].

The integration of DHM tools with VR techniques and Mocap systems improve the level of realism and interaction as demonstrated by case studies described in Sect. 8.3.1. People involved in the experimental tests of the pro-posed VR environments appreciated the interaction control devices and the use of a virtual prototyping environment to validate ergonomically the design of new products. In fact, they considered positively the use of virtual prototypes to perform comparative tests of different design solutions and avoid the use, even if not completely, of physical mock-ups, which are more time consuming and expensive [50].

Previous considerations permit to say that the state of the art of DHM tools makes them applicable within all industrial contexts where virtual prototyping is a technology currently adopted. Company processes that incorporate DHM tools help companies to develop more efficient and safer products that accommodate the needs of a wide range of population. Thus an effective integration of DHM tools with PLM ones will enhance the capabilities of companies to consider and manage ergonomics and human factors since the first stages of the product life cycle saving costs and time [53].

Acknowledgments The author would like to thank in particular Giorgio Colombo, Umberto Cugini, Daniele Movigliatti, and all members of the V&K Group (www.unibg.it/vk) for their contribution to the development of research activities described in this paper.

References

1. Duffy VG (ed) (2007) Digital human modeling. HCI 2007, vol LNCS 4561, Springer, Berlin
2. Sundin A, Ortengren R (2006) Digital human modelling for CAE applications. In: Salvendy G (ed) Handbook of human factors and ergonomics, 3rd edn. Wiley, New York
3. Magnenat-Thalmann N, Thalmann D (eds) (2004) Handbook of virtual humans. Wiley, Chichester
4. Valentini PP (2009) Virtual dummy with spine model for automotive vibrational comfort analysis. Int J Veh Des 51(3/4):261–277
5. Fetter WA (1982) A Progression of human figures simulated by computer graphics. IEEE Comput Graph 2(9):9–13
6. Dooley M (1982) Anthropometric modelling program–a survey. IEEE Comput Graph 2(9):17–25
7. Bonney MC, Case K, Hughes BJ, Kennedy DN, Williams RW (1974) Using SAMMIE for computer-aided workplace and work task design. In: Proceedings of SAE congress, paper 740270. doi: 10.4271/740270
8. http://www.poser.smithmicro.com/poser.html. Accessed February 2011
9. Thalmann D, Grillon H, Maim J, Yersin B, (2009) Challenges in crowd simulation. In: Proceeding 2009 international conference on cyberworlds, Bradford, UK
10. http://www.massivesoftware.com. Accessed February 2011
11. http://crowdit.worldofpolygons.com. Accessed February 2011
12. http://www.diguy.com. Accessed February 2011
13. House DH, Breen DE (eds) (2000) Cloth modeling and Animation. A K Peters Natick, MA, US
14. Seo H, Magnenat-Thalmann N (2003) An automatic modelling of human bodies from sizing parameters. In: Proceedings of the ACM SIGGRAPH symposium on interactive 3D graphics, Monterey, CA, USA
15. Rizzi C, Mana R, Regazzoni D, Cugini U (2006) Virtual apparel, from the customer to the Garment. In: Proceeding of the 16th CIRP international design seminar, design & innovation for sustainable society, Kananaskis, Alberta, Canada
16. Volino P, Luible C, Magnenat-Thalmann N (2007) Virtual clothing–theory and practice. Springer Verlag, Berlin
17. http://www.lectra.com. Accessed February 2011
18. http://www.creasolution.it/cad-tessile-3d-plasma. Accessed February 2011
19. http://www.browzwear.com. Accessed February 2011
20. http://www.optitex.com. Accessed February 2011
21. Wang CCL, Wang Y, Chang TKK, Yuen MMF (2003) Virtual human modeling from photographs for garment industry. Comput Aided Des 35(6):577–589
22. Porter JM, Case K, Freer MT (1996) SAMMIE: a 3D human modelling computer aided ergonomics design system. Co-des J, s2 07 pp 68–75
23. Badler N, Phillips CB, Webber BL (1993) Simulating Humans: computer graphics animation and control. Oxford University Press, Oxford
24. http://www.plm.automation.siemens.com. Accessed February 2011
25. http://www.human-solutions.com. Accessed February 2011
26. http://www.3ds.com/products/delmia/solutions/human-modeling/overview. Accessed February 2011
27. http://www.nexgenergo.com/ergonomics/humancad_prods.html. Accessed February 2011
28. http://www.boeing.com/assocproducts/hms/index.html. Accessed February 2011

29. van der Meulen P, Seidl A (2007) Ramsis–the leading CAD tool for ergonomic analysis of vehicles. In: Duffy VG (ed) Digital human modeling, HCI 2007, vol LNCS 4561. Springer, Berlin Heidelberg
30. Colombo G, Cugini U (2005) Virtual humans and prototypes to evaluate ergonomics and safety. J Eng Des 16(2):195–203
31. Sabbah O, Zaindl A, Bubb H (2009) Design of a mock-up for supported Ingress/Egress using a DHM. In: Proceeding digital human modeling for design and engineering conference and exhibition, Gothenburg, Sweden
32. Berger U, Lepratti R, Otte H (2004) Application of digital human modelling concepts for automotive production. In: Proceeding of the TMCE 2004, Lausanne, Switzerland
33. Mueller A, Maier T (2009) Vehicle layout conception considering vision requirements–a comparative study within manual assembly of automobiles. In: Proceeding of the digital human modeling for design and engineering conference and exhibition, Gothenburg, Sweden
34. Zhang L, Yuan X, Wang L, Dong D (2007) Design and implementation of ergonomics evaluation system of 3D airplane cockpit. In: Duffy VG (ed) Digital human modeling, HCI 2007, vol LNCS 4561. Springer, Berlin
35. Nadadur G, Parkinson M (2009) Using designing for human variability to optimize aircraft eat layout. In: Proceeding digital human modeling for design and engineering conference and exhibition, Gothenburg, Sweden
36. Green RF, Hudson JA (2011) A method for positioning digital human models in airplane passenger seats. In: Duffy VG (ed) Advances in applied digital human modeling. CRC Press, Boca Raton, FL
37. Dantas Alves Silva FR, Rodrigues Miranda M, Reda F (2007) The Impact of the digital human modeling on the aircraft interior projects. In: Proceeding of the digital human modeling conference, Seattle, WA, USA
38. Liem A, Yang H (2004) Digital human models in work system design and simulation. In: Proceeding of the digital human modeling for design and engineering symposium, Rochester, MI, USA
39. Dao TT, Pouletaut F, Marin P, Aufaure P, Charleux F, Ho Ba Tho MC (2010) Simulation of the gait of a patient specific model of post polio residual paralysis (PPRP): effect of the orthosis. In: Proceedings of the 3rd international conference on the development of BME, Vietnam
40. Bucca G, Buzzolato A, Bruni S (2009) A mechatronic device for the rehabilitation of ankle motor function. J Biomech Eng 131(12):7. doi:10.1115/1.4000083
41. Annual Report of the Kessler Medical Rehabilitation Research and Education Center. https://kesslerfoundation.org/research/pdf/innovations/Innovations_8_1.pdf. Accessed February 2011
42. http://www.lifemodeler.com. Accessed February 2011
43. http://www.motionanalysis.com/html/movement/simm. Accessed February 2011
44. http://www.tass-safe.com/madymo_humanmodels. Accessed February 2011
45. http://www.santoshumaninc.com. Accessed February 2011
46. Abdel-Malek K et al (2009) A physics-based digital human model. Int J Veh Des 51 (3/4):324–340
47. Naumann A, Roetting M (2007) Digital human modelling for design and evaluation of human-machine systems. MMI-Interaktiv 12:27–35
48. Bronco D, Rizzi C, Colombo G, and Regazzoni D (2007) Virtual Ergonomics analysis to validate product design. In: Proceeding of the industrial simulation conference 2007. Eurosis publication, Delft, The Netherlands
49. Colombo G, De Angelis F, Formentini L (2010) Integration of virtual reality and haptics to carry out ergonomic tests on virtual control boards. Int J Prod Dev (IJPD) 11(1/2):47–61

50. Bordegoni M, Caruso G, Ferrise F (2008) Mixed-reality environment based on haptic control system for a tractor cabin design review. In: Proceeding of the CIRP design conference 2008, Enschede, The Netherlands
51. Colombo G, De Ponti G, Rizzi C (2010) Ergonomic design of refrigerated display units. Virtual Phys Prototyp 3(5):139–152
52. Colombo G, Facoetti G, Gabbiadini S, Rizzi C (2010) Virtual configuration of lower limb prosthesis. In: Proceedings of the ASME 2010 international mechanical engineering congress & exposition IMECE 2010, Vancouver, British Columbia, Canada
53. Demirel OD, Duffy VG (2007) Digital human modeling for Product life-cycle management. In: Duffy VG (ed) Digital human modeling, HCI 2007, vol LNCS 4561. Springer, Berlin

Chapter 9
CAD and the Rapid Construction of Physical Objects

Stefano Filippi

Abstract This chapter deals with the generation of the physical representation of objects modeled using 3D CAD systems. There are many technologies at the moment that perform this task in a quasi-automatic way. The resulting objects can serve as prototypes in each phase of the design process—in this case the specific term used is rapid prototyping—, or they could be tools for building products—rapid tooling—, or, in some case, they represent the result of the development, the final product—rapid manufacturing—. Then, the role of these technologies appears quite important throughout the whole product development process. They allow resource saving, better product quality, shorter TTM, etc., but at the same time they present specific requirements that inevitably end to influence the product design. Some design for manufacturing methods and tools have been developed and adopted in different fields, in order to help designers and engineers to keep these requirements into consideration. In the following, some technologies will be introduced; then their requirements are exploited to show the data elaboration needed to gain the compatibility with them; finally, some examples of adoption of these technologies in different fields will be reported.

9.1 Introduction

The generation of the physical representation of a product is an important stage of the design and engineering process. Many evaluation activities, e.g., ergonomic and usability checks, manufacturing and assembly verifications, etc., need this kind of representation to be performed at best and to maximize their effectiveness. Moreover, this representation boosts the TTM, increments the flexibility and

S. Filippi (✉)
DIEGM Department, University of Udine, Via delle Scienze, 208, 33100 Udine, Italy
e-mail: filippi@uniud.it

M. Bordegoni and C. Rizzi (eds.), *Innovation in Product Design*,
DOI: 10.1007/978-0-85729-775-4_9, © Springer-Verlag London Limited 2011

decreases the costs of the product development, and helps in improving the product quality [1]. There are specific terms referring to different types of physical object generation, but rapid prototyping—RP—is the generic way used to identify these technologies. All of them are based on material addiction instead of the classic subtractive approach.

The first RP technologies became available in the late 1980s and were used to produce prototype parts [2]. Today, they are used for a much wider range of applications and are even used to manufacture production-quality parts in relatively small numbers (rapid manufacturing), or to build tools used for manufacturing (rapid tooling).

Actually, the term rapid does not refer to the speed in generating the final result; sometimes the generation of an RP object requires much more time than making it using classic, subtractive approaches; instead, it highlights the direct link between the 3D CAD software package used to generate the digital model and the building technology exploited to solidify it.

In some way, the RP could be seen as following the "what you see is what you get—WYSIWYG" paradigm. In fact, the result of the RP process is—or should be—a real copy of the digital model displayed by the 3D CAD software package where the building data come from. Ideally, at the end of the CAD activities, a one-keystroke action should generate the physical representation of the result, without the need of any manual data pre-processing, or the presence of technology experts. Unfortunately, things are not so fair in the real world. CAD data are rarely good enough to be used as-they-are as input for the RP process for many reasons. These reasons will be described in the following.

In this context, another important actor of the design process must be cited, the reverse engineering (RE) [3]. The reason is that, while the RP makes a transformation from the digital representation to the physical one, the RE acts exactly the reverse, by allowing generating the digital model starting from a real, physical object. This RE-CAD-RP closed circle has many interesting implications, as will be highlighted again in the following.

This chapter is organized as follows. The most known and used RP technologies are introduced first, in order to understand their characteristics and requirements. Then, the whole RP process is described, starting from the RP-oriented CAD data generation—in order to sensitize designers' awareness about the issue—up to the selection of the RP technology and the consequent data elaboration. Finally, some examples concerning the RP adoption are listed, in order to show how these technologies can be effectively exploited in heterogeneous application fields.

9.2 RP Technology Overview

There are many RP approaches at the moment, each of them with its own characteristics, in terms of technology used, application domains, working requirements, etc. Fortunately, they share many needs from the CAD point of view, as file formats and data processing. This paragraph describes some of the most used technologies, in order to allow understanding of both the common requirements and the specific ones, time-by-time and technology-by-technology. The RP technologies named stereolitography, direct metal laser sintering, laminated object manufacturing, fused deposition modeling and 3D printing, are described hereafter.

All the RP technologies cited here share the way the object is built-up, the material addiction. Moreover, the result of the RP activity always consists in an ordered set of material layers, stacked one over another, corresponding to the sections of the CAD model sliced at equal intervals on the Z direction. As it will be clear, the layering implies some drawbacks, e.g., the staircase effect and a lack of smoothness on curved surfaces with particular orientations in the RP equipment workspace [4].

9.2.1 Stereolitography

As depicted in Fig. 9.1, in the RP technology called stereolitography (SLA), a mobile platform that can be lifted and lowered is located in the thickness of a layer below the surface of a liquid photo-sensitive polymer contained in a tank. Each section of the model is etched onto the polymer surface by a laser beam that solidifies the polymer almost instantaneously. Once the laser has covered the whole surface of a layer, the platform lowers to a depth of another layer thickness, allowing the liquid resin to flow over the previously cured layer. A recoating blade passes over the surface to ensure that a flat layer of liquid polymer is present, before the etching of the next layer. Supports are required when islands—portions of a layer disconnected from any other portion of the same layer —, overhang, or cantilevered sections exist in the prototype being built [5]. At the end of the building phase, the object is carefully removed from the platform, the liquid resin still present is drained and a post-curing phase is performed in a UV—Ultraviolet—beam oven, to solidify completely the object [6].

Different building styles can be used with SLA. Standard style builds full-resin objects while other styles, for example the quickcast, leave some resin in the liquid state during the slice etching. This is done for different purposes, such as stresses minimization and generation of models for investment casting [7]. SLA prototypes have good dimensional accuracy and surface texture; however, the orientation of the model in the workspace, due to the staircase effect, and the presence of supports can influence the surface finishing [8, 9].

9.2.2 Selective Laser Sintering

The selective laser sintering (SLS) process is one of the most interesting RP technologies currently available. It can build metal or plastic objects using the same material as that of the final product. This characteristic widens the field of application of this technology, which can also be used in rapid tooling, i.e. for the generation of

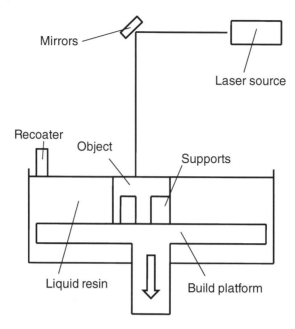

Fig. 9.1 SLA process outline

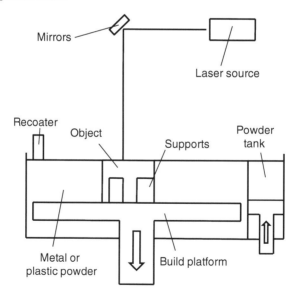

Fig. 9.2 SLS process outline

inserts for plastic injection molding [10, 11], and in rapid manufacturing, to build small series of complex mechanical parts [4, 12]. The object generation process is quite similar to the SLA one, but here there are metal or plastic powders instead of liquid resins. Figure 9.2 shows a mobile platform where a first layer of plastic or

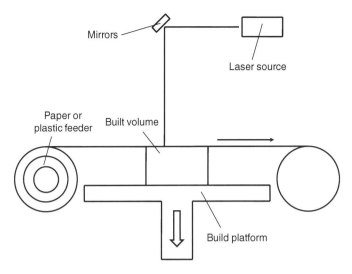

Fig. 9.3 LOM process outline

metal powder collected form a tank is deposed by a recoater. Then the laser beam hatches the powder selectively, and this generates the first solid layer of the object. The platform is lowered, the recoater spreads the next powder layer and the process goes on till the generation of the full object has taken place.

Supports are required even by this technology. Of course, their removal is more difficult than in SLA, especially if metal powders are used. The separation between the object and the platform presents the same problem. Moreover, the SLS process shows some critical aspects mainly due to the behavior of metal powders, such as complex sintering dynamics, residual stress and thermal deformation. [13], so that more study is needed to solve or avoid current limitations even in early activities, for example during the design phase [3].

9.2.3 Laminated Object Manufacturing

The laminated object manufacturing (LOM) exploits rolls of paper or plastic films as material sources to generate the layers. As shown in Fig. 9.3, the left feeder reel provides a new layer, by scrolling the roll toward the right reel. Each new layer is glued to the previous ones, thanks to a polyethylene coating on the bottom side, and etched by a laser beam. This etching draws the contour of the slice coming from the product model, and makes a regular grid on the remaining part of the paper or plastic film surface. At the end of the RP process, the stack of layers appears as a parallele-piped. The physical object is inside a 3D grid that behaves also as a support structure for the overhangs of the model, and that can be easily removed by hand [14].

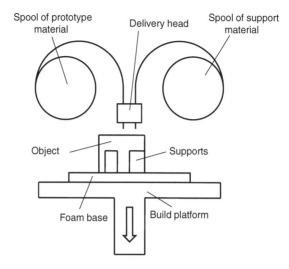

Fig. 9.4 FDM process outline

When all of the wasted material has been removed, the unfinished object is sanded down. This RP technology is mainly used for the generation of large objects, e.g., automotive dashboard or bumper prototypes, because of the building speed and thanks to the cheapness of the process. Regarding the possible problems of this technology, one of them could happen in case of cavities; the removal of the wasted material could be difficult. The paper presents a natural dependence from humidity and temperature, and this can be reduced by coating the model. The surface finishing and the accuracy of the model are not as good as with other RP technologies; however, objects have the look and feel of wood and can be worked and finished like wood.

9.2.4 Fused Deposition Modeling

As described in [15], the fused-deposition modeling (FDM) is an RP technology that builds objects using spools of thermoplastic wire as material. As depicted in Fig. 9.4, the material is heated to just above the melting point in a delivery head. The molten thermoplastic is then extruded through a nozzle as a thin ribbon and deposited in computer-controlled locations appropriate for the layer geometry, thus building a section of the object. Typically, the delivery head moves in the horizontal plane while the foam base, where the object is built, moves vertically, so that each section is built over the previous one. Deposition temperature is such that the deposing material bonds almost instantaneously with that deposited before.

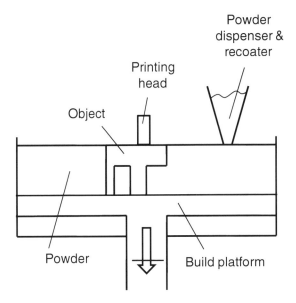

Fig. 9.5 3D Printing process outline

Depending on the geometric complexity of the product model, some supporting material may be necessary to build the object. The amount and the shape of the supports, which will be removed from the final prototype, are automatically calculated, based on the orientation of the product model. The first layer is always built on a grid of supports, slightly larger than the layer, to allow an easy removal of the object from the foam base at the end of the RP process. Precision and surface finishing of the objects are affected by the so-called slicing or layering, which depends on the kind of equipment used. The layer thickness can vary typically from 0.17 mm to 0.33 mm. A wide range of thermoplastic materials can be used to build models, including ABS—Acrylonitrile Butadiene Styrene, polyolefin and polyamide. The final result does not require post-processing, except for support removal and grinding, for a better surface finishing. Another advantage of the FDM is that it can be used not only in a laboratory, but also in an office as well; no high-powered lasers are used and the material, supplied in spool format, does not require special handling or present environmental concerns [2, 4, 16].

9.2.5 3D Printing

The approach to the RP process of the 3D Printing technology is very similar to the ink jet printing process. This is the origin of its name. As shown in Fig. 9.5, a layer of powder, gypsum or other materials, is spread on the built platform by a dispenser. Then, the head "prints" the layer corresponding to the slice of the product model, using glue instead of ink. At the end of a layer, the build platform move downward and the generation of the next layer takes place. The surrounding powder acts as a support volume, so overhangs and isles do not constitute a problem. When the process finishes, the object is carefully extracted from the workspace because of its weakness. Then it can be soaked with resins to gain more hardness and to raise its mechanical properties.

3D Printing is generally faster, more affordable and easier to use than the other RP technologies. Thanks to the printing head, this technology is the only one that allows for the printing of full color prototypes. In fact, it is enough to feed the printing head with different colored glues. Advanced 3D printing technologies yield models that closely emulate the look, feel and functionality of product prototypes. On the contrary, dimensional and geometrical characteristics of the objects made by the 3D Printing are quite scarce [17].

9.3 The Whole RP Process

9.3.1 RP-Oriented CAD Data Generation

Once the 3D CAD model is assumed to be ready for the generation of the corresponding physical object, it must be transferred to the RP equipment. The transfer protocol has two aspects to focus about: the form and the content. The first one may be addressed before the choice of the RP technology, while, for the most part of the second one, the prior selection of a specific RP equipment is mandatory.

The form mainly refers to file formats. The standard *de-facto* is named STL, created by 3D Systems [18]. STL files describe only the surface geometry of a three-dimensional object without any representation of color, texture or other attributes. The STL format presents both ASCII and binary representations. Binary files are more common, since they are more compact. An STL file describes a raw unstructured triangulated surface by the unit normal and vertices of the triangles using a 3D Cartesian coordinate system. The ideal STL file is the one that is ready-to-use, or in other words, if no pre-processing is required to use it as input for the RP equipment and to generate a physical object as an exact representation of the 3D CAD model it represents.

Very often instead, the file is tagged as bad; it presents errors or problems that could drive to the generation of something different from the starting model, or could compromise at all the generation of the physical object. Some of these errors could arise from the form point of view, and in this case they mainly refer to syntax matters; all of this is in charge of software developers, aiming at writing the best translation algorithms for information exchange between 3D CAD software packages and RP equipments.

Problems could arise even from the content point of view, and independently from the RP equipment. It could happen that the STL file syntax is respected and the single triangles are correct, but their set, considered as a whole, presents inconsistencies or have no sense at all. The type of CAD software package used to generate the digital model can make the difference about this. As a matter of fact, a solid modeler generates good STL files, while a geometric modeler could generate inconsistencies in the triangle mesh. The reason is that the STL translation module of a solid modeler can exploit the topological relationships intrinsically present in the boundary representation scheme, while in a geometric modeler every surface is stand-alone; there is no adjacency information. In a solid modeler, a cylinder is a cylinder. In a geometric modeler, a cylinder is a set of three unrelated trimmed surfaces placed in the 3D space. This example helps to understand quite easily why the surface triangulation, needed to generate the STL file, can present inconsistencies. In this case, holes appear in the triangle mesh because the edges of the triangles generated on the planar faces of the cylinder do not match with the edges of those ones generated on the cylindrical surface. Without the adjacency information, the STL translator could make other mistakes; for example, the normal vector of the triangles could be flipped, and this is dramatic for the on-board RP software, because it does not know where are the "in" and the "out", so it does not know where to add material and where not. Fortunately, there are software packages that help a lot in repairing bad STL files because they allow checking and assuring the consistency of the file content. For example, many RP service bureaus use Magics by Materialize [19] to process the STL files before using them. However, an STL file, bad or not, is a simple triangle mesh and the information needed to repair it could be missing. Who can say, for example, that a specific collection of triangles represents a hole, from a functional point of view?

The rest of the STL file content and the possible related problems are discussed after the next paragraph, focused on the selection of the RP technology.

9.3.2 Selection of the RP Technology

When the STL file is reputed clean enough from the syntax point of view, the selection of the RP equipment takes place. This is a fundamental stage because it also determines a further elaboration of data. Given the different characteristics of the RP technologies, there is not one valid for every situation. Then, situations must be characterized at best, in order to make the right choice. About this, the leading criterion is the reason why the physical object is build up. This depends upon the design stage. Starting from the beginning of the product development, a prototype can be considered as conceptual, functional, technical or production sample. This criterion allows setting many requirements, mechanical properties, surface finishing, dimensional and geometric tolerances, delivery times and costs, workspace dimensions, etc., and the comparison between these values and the characteristics of the RP technologies gives helpful hints for the final choice [1]. Table 9.1 qualitatively resumes the correspondence between the type of prototype corresponding to the different design stages and the most suitable technologies for each of them.

Table 9.1 Correspondence between types of prototypes and RP technologies

PROOF-OF	SLA	LOM	SLS	FDM	3D PRINTING
Concept				●	●
Product	●	●		●	
Process		●	●		
Production	●		●		

9.3.3 Model Preparation

Once selected the most suitable RP equipment, the product model must be further processed, in order to match both the design and the technological requirements.

The first action is the setting of the model orientation in the RP equipment workspace. This is a critical decision because it could generate problems related to dimensions and/or geometry, deformations, surface finishing and delivery times and costs. Moreover, the orientation determines the need for supports, for those technologies that require them. Supports are intrinsically related to surface finishing, because their presence does not allow making smooth surfaces, and for this reason the physical objects require post-processing. Also, the orientation is directly linked to surface finishing as well because of the staircase effect. All of this shows how the RP process parameters are very often intertwined, and this makes the RP activities even more complex.

Next, the most of the RP technologies suffer the shrinking phenomenon. Because of this, the dimensions of the product models must be modified accordingly with the technological requirements. Fortunately, the on-board software of the RP equipments manages this matter automatically.

Another important issue to take care about is the workspace exploitation. The RP process is expensive, and an optimized distribution of the physical objects on the build platform can lower the costs.

Finally, the geometry of the supports must be generated, and fortunately this is done again automatically by the on-board software. After that, the slicing of the model takes place. The result is an ordered set of 2D tool paths, used by the RP equipment to drive the laser beam in the SLA and SLS, the printing head in the 3D Printing, the cutting tool in the LOM, and the heating nozzle in the FDM, in generating the physical object [2].

At this moment, the content point of view of the STL files comes back to the stage. Indeed, it is more difficult than the form one to deal with.

Even if everything sounds well—i.e. the STL file content is consistent—, it could happen that the model morphology cannot be reproduced by the specific equipment because of technological limitations. The easiest example is when the workspace of the RP equipment is smaller than the model to build. These situations are more difficult to manage because they are conceptual ones and regard the fundamentals of design and engineering. In fact, designers and engineers cannot neglect technological requirements during their working activities, and the design resulting from their work must satisfy these requirements [1]. Meanwhile, technologies rapidly evolve, making the knowledge related to them hard to be maintained updated. Design for X was born for this matter; it refers to a collection of methods and tools used to keep into consideration technological requirements in a quasi-automatic way, by allowing managing, suggesting, applying and validating design rules, throughout the product development process. Some design for X methods and tools are specifically related to the RP matters; the challenge is to collect the technological information, to formalize it, and to make it available in a usable way [15]. Some 3D CAD software packages start to offer design guidelines regarding manufacturing, assembly, etc. to the users. These guidelines are the concrete results of the design for X.

Another cause of bad STL files from the content point of view, always depending from the chosen RP equipment, could be a wrong setting of the mesh parameters. They control the translation of the free-form surfaces into triangles. High-tolerance values may generate inaccurate, coarse results, while narrow intervals could drive to large files, full of needless information because the represented accuracy is much higher than the technological capabilities of the RP equipment [2].

9.3.4 RP Process Summary

Figure 9.6 summarizes the RP process described up to now.

9.4 RP Applications

RP technologies are widely used in several application fields. This paragraph describes some experiences carried out in heterogeneous contexts, from medical fields to industrial engineering, up to cultural heritage, by the design tools and methods for industrial engineering research group of the University of Udine, now product innovation research group—PIRG—[20], in the last years. Each of them highlights different aspects on the RP adoption.

9.4.1 Medical Fields

RP is very powerful regarding medical and surgical issues. [21] describes the design of a phonatory prosthesis for patients who have undergone total laryngectomy. In this case, the SLS allowed the generation of plastic parts, as shown in Fig. 9.7, and shortened the design activities. [22] reports the development of hip prostheses, where SLS has been used to generate the bone structure to test

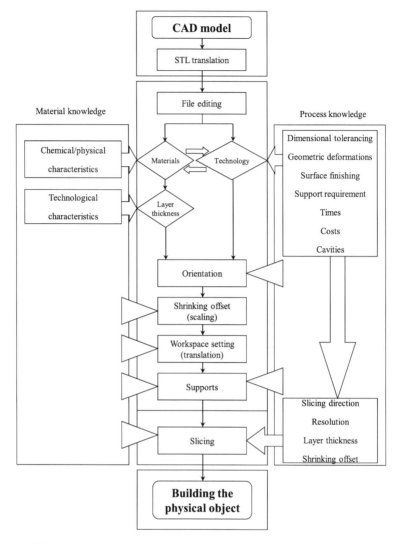

Fig. 9.6 RP process summary

the prosthesis prototype implant (Fig. 9.7). In [23], SLA is used to generate the real copy of the anatomical structure of a patient who needs maxillo-facial surgery. This model is used to simulate the surgical procedure, as shown in Fig. 9.7.

In [24], SLA is exploited to generate the prototype of a knee prosthesis thanks to vacuum casting (Fig. 9.8). Bandera et al. [25] exploited the SLA to validate the results of different RE technologies (Fig. 9.8). In [26], the SLA model is used to validate the results of an innovative method for the development of soft sockets for lower-limb prostheses (Fig. 9.8).

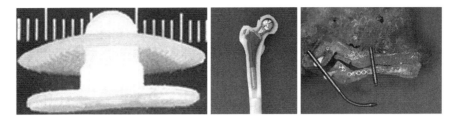

Fig. 9.7 Examples of RP adoption in medical fields: the prototype of a phonatory prosthesis, the bone structure used to test a hip prosthesis implant, and a maxillo-facial surgical activity planning using an RP model

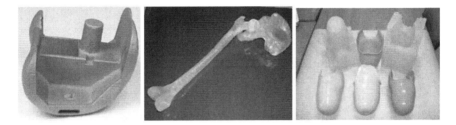

Fig. 9.8 Examples of RP adoption in medical fields: a component of knee prosthesis, a femur and part of a pelvis, and some soft sockets for lower-limb prostheses

9.4.2 Industrial Engineering

The experiences of RP adoption in the industrial engineering field gave an important contribution to the research they were associated to. Many of them allowed collecting the data needed to generate the knowledge base of some design for X methods and tools. This way, these experiences made possible the generation of usable sources of information for other designers and engineers. In [27], SLA has been used to design a component of an industrial garment printer (Fig. 9.9). The matching between the project specifications and the RP requirements started the generation of the knowledge base of a design for manufacturing method named design guidelines—DGL—[28]. Miani et al. reports a procedure to collect data regarding the SLS RP technology [29], while in [30] the roadmap developed thanks to these data helped in developing an innovative model of coffee machine using the SLS (Fig. 9.9). In [31] and [32], some SLS specimens are used to collect data about this RP technology (Fig. 9.9), while in [25] the same action is performed for FDM.

Bandera et al. [33] describes the use of SLS during the design of a mold insert, and the advantages coming from this adoption; for example, the availability of free-form cooling channels (Fig. 9.10). In [34], the rapid tooling offered by the SLS is used to produce the mold for a component of a camera tripod (Fig. 9.10) Bandera et al. [35] describe the use of FDM to validate the knowledge base and the

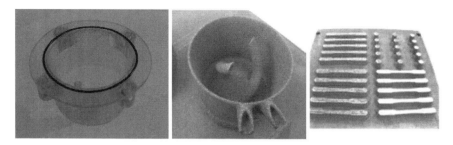

Fig. 9.9 Examples of RP adoption in industrial engineering: a component of a garment printer, a coffee machine, and some SLS specimens still on the build platform

Fig. 9.10 Examples of RP adoption in industrial engineering: a mold insert for plastic injection, the mold of a camera tripod component and a mechanical component in a coordinate measuring machine

effectiveness of the evolution of the DGL, a design for manufacturing and verification tool named DGLs (Fig. 9.10).

In [36] and in [28], the FDM is exploited again to test the validity of the DGLs (Fig. 9.11), while [37] describes the exploitation of FDM to increase the information content of the DGLs-CF, the design for multi-X framework resulting from a further refinement of the previous DGLs [15] (Fig. 9.11).

9.4.3 Cultural Heritage

In [38], SLA has been used for the restoration of pieces of ancient pottery (Fig. 9.12), while in [39] some Celtic fibulae and in [40] some Roman ones have been replicated using the SLA (Fig. 9.12) to generate the model for the vacuum casting process. [41] reports the use of SLA to replicate a small roman bronze statue using vacuum casting. In this case the 3D CAD model first, and the physical afterward, allowed verifying some hypotheses about the real shape of the original statue, given that some parts of it were missing. Figure 9.12 shows the result of this reconstruction.

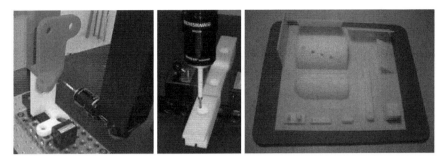

Fig. 9.11 Examples of RP adoption in industrial engineering: two assembled mechanical components in a measuring machine, a mechanical part, and a test bed

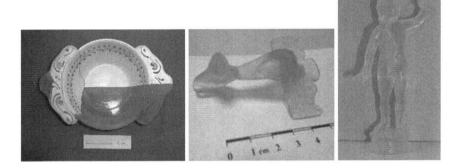

Fig. 9.12 Examples of RP adoption in cultural heritage: pottery restoration, a Celtic fibula, and a small Roman statue

9.4.4 Hybrid Applications

It can happen that the RP is exploited in different contexts at the same time, in order to highlight common elements and design behaviors. The goal is to generate a general roadmap for the RP adoption, customizable time-by-time and case-by-case. In [24], for example, the restoration of some archeological finds and the generation of some tooling for maxillo-facial surgery are put into relationship and a roadmap, starting from data acquisition and ending with the prototype validation is developed. In [42], this roadmap is followed to design and test an impeller of a centrifugal pump. The SLA generates the master model to be used in the investment casting process (Fig. 9.13).

9.5 Conclusion

RP technologies prove to be a powerful tool for product development. Sometimes they are the only way to perform fast and effective evaluations, in order to optimize the product design and manage the resource involved at best. Exactly as it

Fig. 9.13 A centrifugal pump impeller, developed thanks to a SLA master model

happens for classic manufacturing processes, they present requirements that designers and engineers must keep into consideration in their activities, to get the product model compatible with the RP equipment used to solidify it. In doing this, they are supported by design for X methods and tools that make the knowledge about these technologies ready at disposal, in a usable way, as design guidelines. Even if virtual and augmented reality is rapidly growing up in the industrial engineering landscape, they can be seen more complementary than alternative to the RP technologies, because both present different targets. Some research is on the way, focused on a mixed approach, where the prototype is made by a physical component representing the fixed part, and a virtual section projected on the first one, to gain in flexibility and interaction possibilities.

Acknowledgments The author would like to thank Prof. Camillo Bandera, Prof. Massimo Robiony, Dr. Ilaria Cristofolini, Dr. Barbara Motyl and Eng. Daniela Barattin for their precious collaboration in the research activities that generate the results described in this chapter.

References

1. Ullman D (2002) The mechanical design process. McGraw-Hill, New York
2. Pham DT, Dimov SS (2001) Rapid manufacturing. Springer-Verlag, Berlin, Heidelberg, New York
3. Otto K, Wood K (2000) Product design techniques in reverse engineering and new product development. Prentice Hall, New Jersey

4. Chua CK, Leong KF, Lim CS (2003) Rapid prototyping: principles and applications, 2nd edn. World Scientific Publishing Company, Singapore
5. Cheng W, Fuh JYH, Nee AYC, Wong YS, Miyazawa T (1995) Multi objective optimization of part building orientation in stereolithography. Rapid Prototyp J 1(4):12–23
6. Banerjee PS, Sinha A, Banerjee MK (2002) A study on effect of variation of SLA process parameters over strength of Built model. Proceedings of the 2nd national symposium on rapid prototyping and rapid tooling technologies, pp 79–84
7. Chockalingam K, Jawahar N, Ramanathan KN, Banerjee PS (2006) Optimization of stereolithography process parameters for part strength using design of experiments. Int J Adv Manuf Technol 29(1–2):79–88
8. Harris R, Hopkinson N, Newlyn H, Hague R, Dickens P (2002) Layer thickness and draft angle selection for stereolithography injection mould tooling. Int J Prod Res 40(3): 719–729
9. Williams RE, Komara Giri SN, Melton VL, Bishu RR (1996) Investigation of the effect of various build methods on the performance of rapid prototyping (stereolithography). J Mater Process Technol 61(1–2):173–178
10. Kuzman K, Nardin B, Kovac M, Jurkosec B (2001) The integration of Rapid Prototyping and CAE in mould manufacturing. J Mater Process Technol 111:279–285
11. Nelson C (2002) RapidSteel 2.0 Mold Inserts for Plastic Injection Molding © by DTM Technology. http://www.dtm-corp.com. Accessed February 2011
12. Jacobs PF (1995) Stereolithography and other RP&M technologies: from rapid prototyping to rapid tooling. Society of Manufacturing Engineers, Michigan
13. Agarwala M, Bourell D, Beaman J, Marcus H, Barlow J (1995) Direct selective laser sintering of metals. Rapid Prototyp J 1(1):26–36
14. Kechagias J (2007) Investigation of LOM process quality using design of experiments approach. Rapid Prototyp J 13(5):316–323
15. Filippi S, Cristofolini I (2009) The design guidelines collaborative framework: a design for Multi-X method for product development. Springer, Berlin
16. Cooper K (2001) Rapid prototyping technology: selection and application. Taylor and Francis Group Ltd, London
17. Bassoli E, Gatto A, Iuliano L et al (2007) 3D printing technique applied to rapid casting. Rapid Prototyp J 13(3):148–155
18. Chua CK, Gan JGK, Tong M (1997) Interface between CAD and rapid prototyping systems.1. A study of existing interfaces. Int J Adv Manuf Technol 13(8):566–570
19. http://www.materialise.com/Magics. Accessed February 2011
20. http://www.diegm.uniud.it/PIRG/. Accessed February 2011
21. Miani C, Bergamin AM, Staffieri A, Filippi S, Miani F, Zanzero M (1999) Experiences in rapid prototyping: voice devices for patients who have undergone total laringectomy. In: Kuljanic E (ed) Advanced manufacturing systems and technology, vol 406., CISM courses and lecturesSpringer-Verlag, Wien, New York
22. Filippi S, Bandera C, Felice M (2001) Cooperative Work in medicine: linking together distributed expertise in hip-prostheses development and surgical planning. Proceedings of ISPE-CE2001, Advances in CONCURRENT ENGINEERING, California, USA
23. Robiony M, Salvo I, Costa F, Zerman N, Bandera C, Filippi S, Felice M, Politi M (2008) Accuracy of virtual reality and stereolithographic models in maxillo-facial surgical planning. J Craniofacial Surgery 19(2):482–489
24. Bandera C, Filippi S, Motyl B (2004) Merging design activities among different application fields: from medicine and cultural heritage to industrial engineering. Proceedings of the Design 2004, 8th international design conference, Dubrovnik, Croatia
25. Bandera C, Cristofolini I, Filippi S (2005) Customising a Knowledge-Based System for design optimisation in Fused Deposition Modelling RP-technique. In: Kuljanic E (ed) Advanced manufacturing systems and technology, vol 486., CISM courses and lecturesSpringer-Verlag, Wien New York

26. Colombo G, Filippi S, Rizzi C, Rotini F (2008) A Computer assisted methodologies to improve prosthesis development process. Proceedings of CIRP design conference 2008, Twente (The Nederlands)
27. Filippi S, Bandera C, Felice M (2000) Formalizzazione della conoscenza e progettazione basata su prototipo: esempi di applicazione. Proceedings of III Seminario de Bilbao, Spain (in italian)
28. Filippi S, Cristofolini I (2007) The Design Guidelines (DGLs), a Knowledge Based System for industrial design developed accordingly to ISO-GPS (Geometrical Product Specifications) concepts. Research in Engineering Design, Springer, London, 18(1):1–19
29. Miani F, Kuljanic E, Filippi S, Guggia R (2002) On some developments of direct metal selective laser sintering. Proceedings of MIT 2001 5th slovenian conference on management of innovative technologies, Fiesa (Slovenia)
30. Filippi S, Bandera C, Toneatto G (2001) Generation and testing of guidelines for effective rapid prototyping activities. Proceedings of ADM international conference on design tools and methods in industrial engineering, Rimini, Italy
31. Bandera C, Filippi S, Miani F, Toneatto G (2002) Exploiting the evaluation of RP artefacts to update knowledge in design guidelines. In: Kuljanic E (ed) Advanced manufacturing systems and technology, vol 437., CISM courses and lecturesSpringer, Wien New York, pp 741–747
32. Filippi S, Gasparetto A, Miani F (2002) Study of the dynamic response of a sinterized material to a seismic excitation for knowledge enhancement in an expert system for rapid prototyping. Proceedings of MOVIC '02, Japan
33. Bandera C, Filippi S, Toneatto G (2003) Progettazione basata sulla conoscenza di un inserto per stampo industriale realizzato con tecnologia DMLS. Proceedings of XIII ADM-XV INGEGRAF international conference on tools and methods evolution in engineering design, Napoli, Italy (in italian)
34. Bandera C, Filippi S, Toneatto G, Corrent E, De Cet R (2004) Rapid tooling per la presso-fusione di leghe di alluminio in uno scenario di progettazione collaborativa: un'esperienza industriale. Proceedings of XIV ADM–XXXIII AIAS: Innovazione nella Progettazione Industriale, Bari, Italy (in italian)
35. Bandera C, Cristofolini I, Filippi S (2006) Using a knowledge-based system to link design, manufacturing, and verification in a collaborative environment. Proceedings of CIRP ICME 06–5th CIRP international seminar on intelligent computation in manufacturing engineering, Ischia, Italy
36. Bandera C, Cristofolini I, Filippi S (2007) Design for X: a roadmap to apply a knowledge-based system for product optimization in a collaborative environment. Proceedings. of XVI ADM–XIX INGEGRAF, Perugia, Italy
37. Bandera C, Cristofolini I, Filippi S, Motyl B (2008) Evaluation of FDM performance to enhance a collaborative framework for product redesign and process reconfiguration. In: Kuljanic E (ed) Advanced manufacturing systems and technology CISM courses and lectures, pp 339–349
38. Bandera C, Filippi S, Motyl B (2002) Computer-supported-cooperative-work strategies in cultural heritage preservation. Proceedings of eurographics '02, Milano, Italy
39. Bandera C, Felice M, Filippi S (2002) CT-based reverse engineering and rapid prototyping: experiences in different application domains. In: Kuljanic E (ed) Advanced manufacturing systems and technology, vol 437., CISM courses and lecturesSpringer-Verlag, Wien New York, pp 741–747
40. Bandera C, Filippi S, Motyl B (2003) Remaking of roman fibulae in a CSCW environment based on reverse engineering and rapid prototyping. EUROGRAPHICS 2003–Italian Chapter, Milano, Italy
41. Bandera C, Filippi S, Motyl B (2003) Acquisizione, prototipazione e replica di un reperto archeologico: il bronzetto di Zuglio. Proceedings of XIII ADM-XV INGEGRAF international conference on tools and methods evolution in engineering design, Napoli, Italy (in italian)
42. Bandera C, Filippi S, Motyl B (2006) Validating CSCW strategies and applications for rapid product development in investment casting process. IJPR Int J Prod Res 44(9):1659–1680

Subject Index

M. Bordegoni and C. Rizzi (eds.), *Innovation in Product Design*,
DOI: 10.1007/978-0-85729-775-4, © Springer-Verlag London Limited 2011

Printed by Publishers' Graphics LLC USA
MO20120323-006
2012